Advanced Lectures in Mathematics

Variational Principles for Discrete Surfaces

Theories and Algorithms

Feng Luo · Xianfeng David Gu · Junfei Dai

高等教育出版社
HIGHER EDUCATION PRESS

International Press

图书在版编目 (CIP) 数据

离散曲面的变分原理=Variational Principles for Discrete Surfaces: 英文/罗锋, 顾险峰, 戴俊飞著. —北京 : 高等教育出版社, 2008.1

ISBN 978-7-04-023194-6

I. 离··· II. ①罗··· ②顾··· ③戴··· III. 多面体—英文

IV. 0189

中国版本图书馆 CIP 数据核字 (2007) 第 176547 号

策划编辑 王丽萍 **责任编辑** 王丽萍
封面设计 张申申 **责任印制** 陈伟光

出版发行	高等教育出版社	**购书热线**	010–58581118
社　　址	北京市西城区德外大街 4 号	**免费咨询**	800–810–0598
邮政编码	100011	**网　　址**	http://www.hep.edu.cn
总　　机	010–58581000		http://www.hep.com.cn
		网上订购	http://www.landraco.com
经　　销	蓝色畅想图书发行有限公司		http://www.landraco.com.cn
印　　刷	涿州市京南印刷厂	**畅想教育**	http://www.widedu.com
开　　本	787×1092　1/16	**版　　次**	2008 年 1 月第 1 版
印　　张	9	**印　　次**	2008 年 1 月第 1 次印刷
字　　数	165 000	**定　　价**	26.00 元

本书如有缺页、倒页、脱页等质量问题, 请到所购图书销售部门联系调换。

物 料 号　23194–00

ADVANCED LECTURES IN MATHEMATICS

Dedicated to our parents, wives and kids.

Preface

This book consists of mathematical and algorithmic studies of geometry of polyhedral surfaces based on the variations principles. The part of mathematics is based on a lecture series given by Feng Luo at the Center of Mathematical sciences at Zhejiang University, China, in June and July 2006. The algorithmic theory and applications to computer graphic are based on the work of Xianfeng Gu and are written by him. The task of writing the part of mathematics of the note was done by Junfei Dai who prepared them with great care and made a number of improvements in the exposition.

The aim of this book is to introduce to the students and researchers an emerging field of polyhedral surface geometry and computer graphics based on variation principles. These variational principles are derived from the derivatives of the cosine law for triangles. From mathematical point of view, one of the most fascinating identity in low-dimensional polyhedral geometry is the Schlaefli formula. It relates in a simple and elegant to way the volume, edge lengths and dihedral angles of tetrahedra in the spheres and hyperbolic spaces in dimension 3. The formula can be considered as a foundation of 3-dimensional variational principles for triangulated objects. For a long time, mathematicians have been considering the Gauss-Bonnet formula as the 2-dimensional counterpart of Schlaefli. The recent breakthrough in this area was due to the work of Colin de Verdiere in 1995 who found the first 2-dimensional identity relating edge lengths and inner angles similar to the Schlaefli identity. The mathematical work produced in this book can be considered as establishing all 2-dimensional counterparts of Schaefli formula. It turns out there are continuous families of Schlaefli type identities in dimension 2. These identities produce many interesting variational principles for polyhedral surfaces. In the part of mathematics of the book, we are focusing on a study of the rigidity phenomena on polyhedral surfaces. Some moduli space problems are also discussed in the book.

In the part of algorithm of the book, we introduce discrete curvature flow from both theoretical and practical points of view. Discrete curvature flow is a powerful tool for designing metrics by prescribed curvatures. The algorithm maps general surfaces with arbitrary topologies to three canonical spaces. Therefore, all geometric problems of surfaces in 3D space are converted to 2D ones. This greatly improves the efficiency and accuracy for engineering applications. The discrete Ricci flow algorithm, and Ricci energy optimization algorithm are rigorous, robust, flexible and efficient. They have been applied for surface matching, registration, shape classification, shape analysis and many fundamental applications in practice.

This book is written for senior undergraduate students and graduate students majoring in mathematics or computer science. The mathematical requirements to follow the proofs in the book are some basic knowledge of differential geometry and elementary surface topology. We have not stated and proved theorems in the book in the most general form to avoid the technical details. For computer science majors with basic knowledge in data structure and algorithm, all the algorithms in the book can be implemented straightforwardly by following the pseudo codes and the software system can be built step by step from scratch. Some data sets and source code are also available upon request.

This book should be valuable for researchers in surface geometry, computer graphics, computer vision, geometric modelling, visualization, medical imaging and scientific computation fields. The computational algorithms are also useful for geometric modelers, industrial products designer, digital artists, animators, game developers and anyone who needs digital geometry processing tools.

We owe much to many colleagues and friends with whom we have discussed the subject matter over the years. The first author would like to thank Ben Chow who introduced him the wonderful world of discrete curvature flows and the Ricci flow. The second author is very grateful to all professors in the *Center of Visual Computing* at Stony Brook: Arie Kaufman, Hong Qin, Dimitris Samaras, Klaus Mueller and all faculty members in the Computer Science department in Stony Brook University. The second author deeply appreciate the encouragements and valuable advices from Professor Shing-Tung Yau.

The authors thank National Science Foundation of USA for supporting our collaborative research on discrete curvature flow.

Last but not least, we also want to thank our families, without their supports, this book could not be accomplished.

Piscataway, New Jersey, *Feng Luo*
Stony Brook, New York, *Xianfeng David Gu*
Hangzhou, China, *Junfei Dai*
Summer 2007

Contents

1

Introduction

1.1 Variational Principle and Isoperimetric Problems

The variational approach to characterize a geometric object is to find among geometric shapes of similar properties the one with the least measurement (the least energy). Let us illustrate this theme by the famous isoperimetric problem which asks for a characterization of the circle among all piecewise smooth non-self intersecting loops in the plane. Let $\gamma: S^1 \to \mathbb{C}$ or $\gamma: [0,2\pi] \to \mathbb{C}$ be a piecewise smooth embedding so that $\gamma(0) = \gamma(2\pi)$. Then the length $l(\gamma)$ of γ and the area $A(\gamma)$ enclosed by γ are given by

$$l(\gamma) = \int_0^{2\pi} \sqrt{x'(t)^2 + y'(t)^2 dt}$$

and

$$A(\gamma) = \int\int_{\Omega} dxdy = \oint_{\gamma} xdy = \int_0^{2\pi} x(t)y'(t)dt$$

where $\gamma(t) = (x(t), y(t))$ and Ω is the Jordan domain enclosed by γ. Define the energy of γ, denoted by $E(\gamma)$, by

$$E(\gamma) = l^2/A.$$

Then the isoperimetric inequality (see [Cha] for a proof) states that

$$E(\gamma) \geqslant 4\pi^2\rho^2/\pi\rho^2 = 4\pi^2$$

with equality if only if γ is a round circle.

The variational principle says that circles are the minimizers of energy E among all piecewise smooth Jordan curves. Technically, the difficult part is to prove the existence of minimizes. As long as one knows the existence, it is not hard to see that the minimizer has to be a circle. Indeed, both length l and area A are invariant under rigid motion. It follows that minimizer has to have the rotational symmetry. Thus, it must be a circle.

A counterpart of the isoperimetric problem is that the regular n-sided polygons (or n-gons) are the $E(\gamma)$ energy minimizer among all n-gons. This was first proved by Steinitz. The argument is elementary and elegant. First, due to the finite dimensionality, one can show that the energy minimizer exists. Next, one shows that the minimizer must be convex due to the reflection operation which increases the energy (see Fig. 1.1).

It is known that among all triangles of area 1 with the same base, the isosceles triangle has the smallest circum length (see Fig. 1.2). Using this fact to the minimizer, one sees that the edge lengths of the minimizer must be the same. Finally, Steinitz showed that the

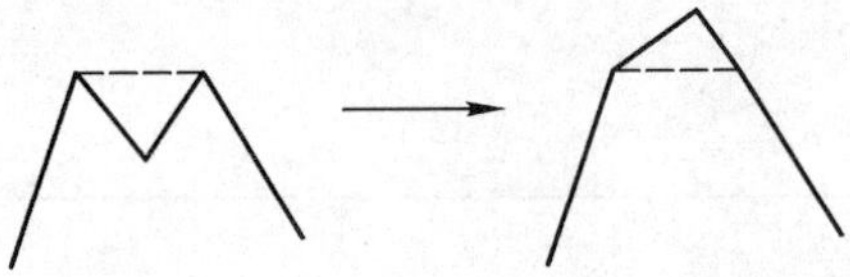

Fig. 1.1. Reflection

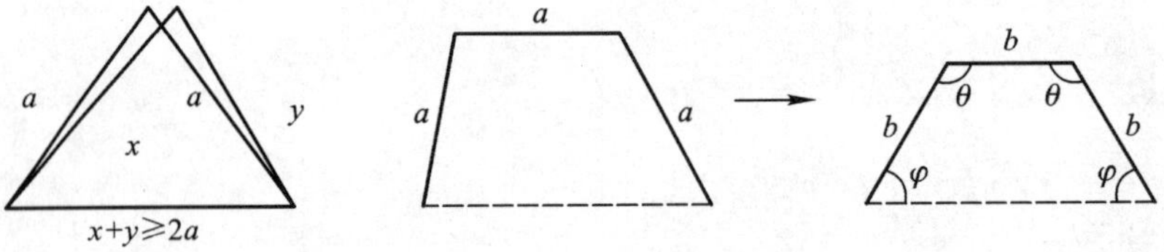

Fig. 1.2.

minimizer must be regular by establishing the fact that among the quadrilaterals of area 1 with the same base, the trapezoid of the same top two angles has the smallest circumlength.

The Steinitz's solution above is an example of variational principle on triangulated 1-manifolds.

The main mathematical goal of our study in this book is to consider the 2-dimensional analogous of the Steinitz's solution. We will consider triangulated surfaces and construct the associated energy functions. Among the main results obtained in this work is the complete list of all possible energy functions of localized nature for 2-dimensional variational principles on triangulated surfaces.

1.2 Polyhedral Metrics and Polyhedral Surfaces

Take a finite set of points in the Euclidean 3-space $\mathbb{E}^3$ and the convex hull of these points, we obtain a convex polytope whose vertices are among the given finite set. If the points are generic, then the convex polytope has triangle faces. In this case, the boundary surface is a polyhedral surface. It has two properties. Firstly the surface is triangulated, secondly, the induced metric on the surface is locally flat except at the vertices. Recall that a triangulation of a surface is defined as follows. Take a finite collection of disjoint triangles and identify all pairs of edges by homeomorphisms. The quotient space is a surface with a triangulation whose cells are the quotients of triangles, edges and vertices in the disjoint union.

Definition 1.1. *An Euclidean polyhedral surface is a triple (S,T,d) where S is a closed surface, T is a triangulation of S and d is a metric on S whose restriction to each triangle is isometric to an Euclidean triangle. We will call the metric d a polyhedral metric. The discrete curvature k_0 of (S,T,d) is a function which assigns each vertex 2π less the sum of inner angles at the vertex, i.e.,*

$$k_0(v) = 2\pi - \sum_{i=1}^{m} \alpha_i,$$

where $\alpha_1, ..., \alpha_m$ are the inner angles (of triangles) at the vertex v.

We will also call above an $\mathbb{E}^2$ polyhedral surface. If we use the spherical (or hyperbolic) triangles instead of Euclidean triangles in Definition 1.1, the result is called a spherical (or $\mathbb{S}^2$) polyhedral surface (resp. $\mathbb{H}^2$ polyhedral surface). Spherical and hyperbolic polyhedral surfaces will be covered extensively in this book. It is due to the fact the $\mathbb{S}^2$ and $\mathbb{H}^2$ polyhedral surfaces constitute generic phenomena in polyhedral geometry so that Euclidean ones are the limiting cases. The discrete curvature k_0 is defined by the same formula for $\mathbb{S}^2$ and $\mathbb{H}^2$ polyhedral surfaces.

For a triangulated surface (S,T), we use $E = E(T)$ and $V = V(T)$ to denote respectively the sets of all edges and vertices of the triangulation. A polyhedral metric d on (S,T) is determined by specifying the lengths of edges, i.e., by the edge length function $l_d : E \to \mathbb{R}_{>0}$. In the sequel, we will identify a polyhedral metric d with its edge length function. In particular, we say two polyhedral metrics d, d', are equivalent if they have the same edge length function, i.e., $l_d = l_{d'}$.

From the definition, it is clear that the basic unit of discrete curvature is the inner angle. Furthermore, the metric-curvature relation is given by the cosine law.

Just like the smooth case, one of the main problem of study in polyhedral surface is to understand the relationship between the metric and curvature. Naturally we should study the cosine law carefully.

1.3 A Brief History on Geometry of Polyhedral Surface

The study of polyhedral geometry has a long history. The Greeks were able to find the five regular polytopes in $\mathbb{E}^3$ long time ago. Modern developments started with Euler's famous theorem concerning the number of vertices, edges and faces of convex polytopes in $\mathbb{E}^3$. Additional ideas and results came form Legendre, Cauchy, Steiner, Sylvester, Cayley, Möbius, Schlafli, Tait, Minkovski, Steinitz, and others.

We just mention a few remarkable theorems concerning rigidity and existence of the convex polyhedral surfaces here.

Theorem 1.2 (Cauchy, 1813). *Suppose P,Q are two compact convex polytopes in $\mathbb{E}^3$ so that there exists an isometry $f : \partial P \to \partial Q$ between their boundary surfaces preserving vertices, edges and facess. Then P and Q differ by a rigid motion of $\mathbb{E}^3$.*

Cauchy's original proof contained a gap which was fixed by Steinitz. A 3-connected graph is a graph with the property that if one removes any two vertices, the remaining subgraph is still connected.

Theorem 1.3 (Steinitz, 1917). ([Gru]) *For any 3-connected graph G in the 2-sphere $\mathbb{S}^2$, there exists a compact convex polytope P in $\mathbb{E}^3$ whose edge graph is isomorphism to G by a homeomorphism from ∂P to $\mathbb{S}^2$.*

A very important problem is to understand the space of all compact convex polytopes with combinatorics G. This is called the realization space. Steinitz proved that the space is connected. Conjecturally the space should be a cell.

Theorem 1.4 (Aleksandrov, 1942). *If (S^2,T,d) is a Euclidean polyhedral surface with curvature $k_0 \geqslant 0$, then there exists a compact convex polytope in $\mathbb{E}^3$ whose boundary is isometric to (S^2,d).*

Furthermore, the convex polytope in Aleksandrov's theorem is unique up to isometry of $\mathbb{E}^3$. Note, the uniqueness does not follow directly from Cauchy's rigidity theorem which assumes that boundary isometry preserves combinators. The most general rigidity theorem was established by Aleksandrov and Pogorelov.

Theorem 1.5 (Aleksandrov and Pogorelov, 1948). *If P and Q are two compact convex bodies in $\mathbb{E}^3$ so that their boundaries are isometric, they differ by a rigid motion of $\mathbb{E}^3$.*

Combining Aleksandrov realization theorem and the rigidity theorem, one sees that for a polyhedral surface (S^2,T,d) of non-negation discrete curvature $k_0 \geqslant 0$, there exists a natural cell decomposition T_1 of S^2 coming from the convex polytope realization. On the other hand, for the singular flat metric (S^2,d), there exists a natural Delaunay cell-decomposition T_2 of S^2. It is known that T_1 and T_2 may be different. It is conjecture that T_1 and T_2 differ by a bounded number of flips which is linear in the number of vertices in T_1 (or T_2) (see Fig. 1.3). To be more precise, let N be the number of vertices in T_1. By adding diagonals to non-triangular faces of T_1 and T_2 (without introducing extra vertices), we obtain two triangulations $\widetilde{T}_1$ and $\widetilde{T}_2$. Conjecturally, $\widetilde{T}_1$ and $\widetilde{T}_2$ differ by a finite number of flip operations so that the number of flips is linear in N.

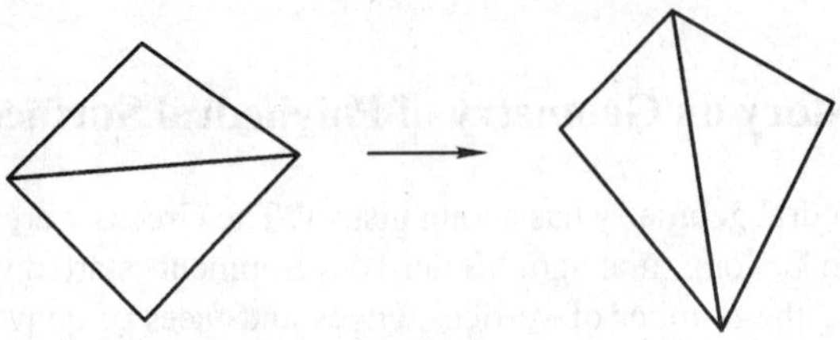

Fig. 1.3.

1.4 Recent Works on Polyhedral Surfaces

The recent insurgent of interests on polyhedral surfaces is mainly due to the work of William Thurston on geometrization of 3-manifolds. Recall that a polyhedral metric (S,T,d) is called a circle packing metric if there exists a positive function $r: V \to \mathbb{R}_{>0}$ so that the edge length l_d of an edge uv with end points u and v is given by

$$l_d(u,v) = r(u) + r(v).$$

For instance a triangle $\Delta v_1v_2v_3$ in T is realized by an Euclidean triangle of edge lengths $r_1+r_2, r_2+r_3, r_3+r_1$. Geometrically, the vertices v_1, v_2, v_3 are the centers of three pairwise tangent circles of radiu r_1, r_2 and r_3 (see Fig. 1.4).

Theorem 1.6 (Thurston[Thu], Andreev[An]). *Suppose (S,T) is a closed triangulated surface. Then*

(a) *An Euclidean circle packing metric on (S,T) is determined up to scaling by its discrete curvature k_0.*

(b) *A hyperbolic circle packing metric on (S,T) is determined by its discrete curvature k_0.*

Furthermore, the space of all discrete curvatures $\{k_0\}$ of $\mathbb{E}^2$ (or $\mathbb{H}^2$) circle packing metric on (S,T) is a convex polytope.

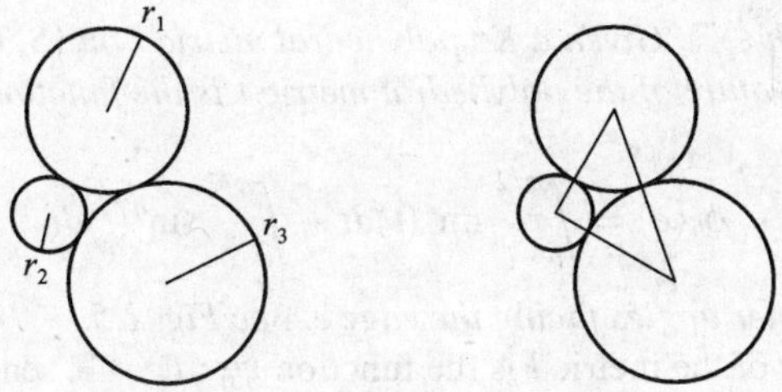

Fig. 1.4.

Definition 1.7 (Leibon[Le]). *Suppose (S,T,d) is a K^2 polyhedral surface where $K^2 = \mathbb{E}^2$ or $\mathbb{S}^2$ or $\mathbb{H}^2$. Then the ψ_0-curvature of it is the map $\psi_0 : \{edges\ in\ T\} \to \mathbb{R}$ so that $\psi_0(e) = \frac{1}{2}(b+b'+c+c'-a-a')$ where a, a' are the angles facing the edge e and b, b', c, c' are the angles adjacent to e.*

Note that for Euclidean polyhedral surfaces, $b+c = \pi - a$ and $b'+c' = \pi - a - a'$. Thus, $\psi_0(e) = \pi - a - a'$.

Definition 1.8. *A polyhedral surface (S,T,d) is called Delaunay if $\psi_0(e) \geqslant 0$ for all edges e in the triangulation. In the case that the discrete curvature $k_0(v) \equiv 0$ at all vertices, i.e, the polyhedral surface (S,T,d) is the geometric triangulation of a constant curvature surface, the Delaunay condition is the some as the classical no-vertex-circle condition. Namely, $\psi_0(e) \geqslant 0$ for all e if and only if the circum-circle of each triangle contains no vertex in its interior (in the universal cover).*

Two remarkable rigidity theorems for Euclidean and hyperbolic polyhedral surfaces are:

Theorem 1.9 (Rivin, 1995). *Any Euclidean polyhedral metric on a closed triangulated surface (S,T) is determined up to scaling by its ϕ_0-curvature $\phi_0 : E \to \mathbb{R}$ so that $\phi_0(e) = \pi - a - a'$ where a, a' are the angles facing e. Furthermore, the set of all ϕ_0-curvatures $\{\phi_0\}$ of all Delaunay Euclidean polyhedral metrics on (S,T) is a convex polytope.*

Note by the remarks above, we have $\phi_0(e) = \psi_0(e)$ in the Euclidean case. Thus Delaunay means $\phi_0 \geqslant 0$.

Theorem 1.10 (Leibon, 2002). *A hyperbolic polyhedral metric on a triangulated closed surface (S,T) is determined by its ψ_0-curvature. Furthermore, the set of all ψ_0-curvatures $\{\psi_0\}$ of Delaunay hyperbolic metrics on (S,T) is a convex polytope.*

1.5 Some of Our Results

Following [Lu1], we introduce three families of curvatures. The relationships between the polyhedral metrics and these curvatures like quantities are the main focus of the study in this book.

Suppose (S,T) is a closed triangulated surface so that T is the triangulation, E and V are the sets of all edges and vertices respectively. Let $\mathbb{E}^2$, $\mathbb{S}^2$ and $\mathbb{H}^2$ be the Euclidean, the spherical and the hyperbolic 2-dimensional geometries recpectily.

Definition 1.11. *Let $h \in \mathbb{R}$. Given a K^2 polyhedral metric l on (S,T) where $K^2 = \mathbb{E}^2$, or $\mathbb{S}^2$ or $\mathbb{H}^2$, the ϕ_h-curvature of the polyhedral metric l is the function $\phi_h : E \to \mathbb{R}$ sending an edge e to:*

$$\phi_h(e) = \int_a^{\pi/2} \sin^h(t)dt + \int_{a'}^{\pi/2} \sin^h(t)dt \tag{1.1}$$

where a, a' are the inner angles facing the edge e. See Fig. 1.5.

The ψ_h-curvature of the metric l is the function $\psi_h : E \to \mathbb{R}$ sending an edge e to

$$\psi_h(e) = \int_0^{\frac{b+c-a}{2}} \cos^h(t)dt + \int_0^{\frac{b'+c'-a'}{2}} \cos^h(t)dt \tag{1.2}$$

where b, b', c, c' are the inner angles adjacent to the edge e and a, a' are the angles facing the edge e. See Fig. 1.5.

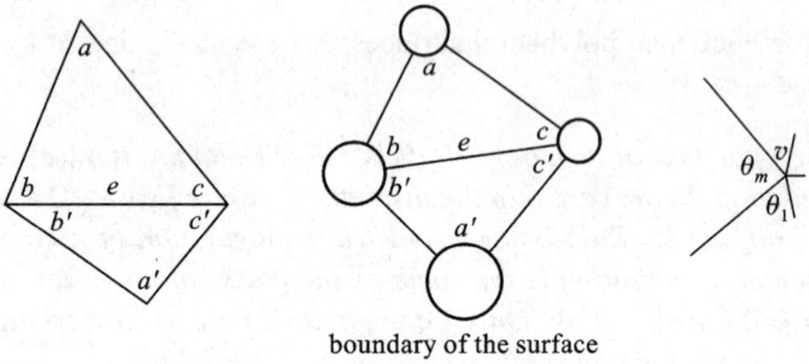

Fig. 1.5.

The h-th discrete curvature k_h of the polyhedral metric l on (S,T) is the function $k_h : V \to \mathbb{R}$ sending a vertex v to

$$k_h(v) = \frac{4-m}{2}\pi + \sum_{i=1}^{m} \int_{\theta_i}^{\pi/2} \tan^h(t/2)dt \tag{1.3}$$

where $\theta_1, \cdots, \theta_m$ are all inner angles at the vertex v. See Fig. 1.5.

We remark that ϕ_0- and ψ_0-curvatures were first introduced by I. Rivin [Ri] and G. Leibon [Le] respectively.

We have

Theorem 1.12 ([Lu1]). *Let $h \in \mathbb{R}$ and (S,T) be a closed triangulated surface.*

(1) *A Euclidean circle packing metric on (S,T) is determined up to isometry and scaling by its k_h-th discrete curvature.*
(2) *A hyperbolic circle packing metric on (S,T) is determined up to isometry by its k_h-th discrete curvature.*
(3) *If $h \leqslant -1$, a Euclidean polyhedral metric on (S,T) is determined up to isometry and scaling by its ϕ_h-curvature.*
(4) *If $h \leqslant -1$ or $h \geqslant 0$, a spherical polyhedral metric on (S,T) is determined up to isometry by its ϕ_h-curvature.*
(5) *If $h \leqslant -1$ or $h \geqslant 0$, a hyperbolic polyhedral surface is determined up to isometry by its ψ_h-curvature.*

For any $h \in \mathbb{R}$, there are local rigidity theorems in cases (3), (4), (5). We believe that above theorem holds for all $h \in \mathbb{R}$.

The counterpart of Theorem 1.12(5) for hyperbolic metrics with totally geodesic boundary on an ideal triangulated compact surface is the following. Recall that an *ideal triangulated compact surface* with boundary (S,T) is obtained by removing a small open regular neighborhood of the vertices of a triangulation of a closed surface. The *edges* of an ideal triangulation T correspond bijectively to the edges of the triangulation of the closed surface. Given a hyperbolic metric l with geodesic boundary on an ideal triangulated surface (S,T), there is a unique geometric ideal triangulation T^* isotopic to T so that all edges are geodesics orthogonal to the boundary. The edges in T^* decompose the surface into hyperbolic right-angled hexagons. The *ψ_h-curvature* of the hyperbolic metric l is defined to be the map $\psi_h : \{$ all edges in $T\} \to \mathbb{R}$ sending each edge e to

$$\psi_h(e) = \int_0^{\frac{b+c-a}{2}} \cosh^h(t)dt + \int_0^{\frac{b'+c'-a'}{2}} \cosh^h(t)dt \tag{1.4}$$

where a, a' are lengths of arcs in the boundary (in the ideal triangulation T^*) facing the edge and b, b', c, c', are the lengths of arcs in the boundary adjacent to the edge so that a, b, c lie in a hexagon. See Fig. 1.5.

Theorem 1.13 ([Lu1]). *A hyperbolic metric with totally geodesic boundary on an ideal triangulated compact surface is determined up to isometry by its ψ_h-curvature. Furthermore, if $h \geqslant 0$, then the set of all ψ_h-curvatures on a fixed ideal triangulated surface is an explicit open convex polytope P_h in a Euclidean space so that $P_h = P_0$.*

The case when $h < 0$ has been recently established by Ren Guo [Gu1]. He proved

Theorem 1.14 ([Gu1]). *Under the same assumption as in theorem 1.13, if $h < 0$, the set of all ψ_h-edge invariants on a fixed ideal triangulated surface is an explicit bounded open convex polytope P_h in a Euclidean space. Furthermore, if $h < h'$, then $P_h \subset P_{h'}$.*

The interesting part of Theorem 1.13 is that the images of the Teichmüller space in these coordinates (for $h \geqslant 0$) are all the same. Whether these coordinates are related to quantum Teichmüller theory is an interesting topic. See [FC, Ka, BL, Te] for more information. Combining Theorem 1.13 with the work of Ushijima [Us] and Kojima [Ko], one obtains for each $h \geqslant 0$ a cell decomposition of the Teichmüller space invariant under the action of the mapping class group.

The space of all polyhedral surfaces will also be studied in this book. In particular, we will give a Marden-Rodin's proof of Thurston's theorem that the space of all discrete curvatures $\{k_0\}$ of hyperbolic circle packing metrics on (S,T) is a convex polytope.

1.6 The Method of Proofs and Related Works

The variational principles were used to the proofs of above theorems. The use of variational principle on triangulated surfaces recently appeared in the seminal work of Colin de Verdiere [CV1] in 1991. In this work, Colin de Verdiere gave a new proof of Thurston-Andreev's circle packing theorem using a variational principle. The most striking part of the proof in [CV1] is the discovery of an action functional for the geometric triangles. This action functional of Colin de Verdiere can be considered as a 2-dimensional analogous of the Schlaefli formula. Recall that for a tetrahedron in $\mathbb{H}^3$ or $\mathbb{S}^3$, if l_e and θ_e are the length and dihedral angle at the edge e, then Schlaefli formula says

$$\frac{\partial V}{\partial \theta_e} = \frac{\lambda}{2} l_e \tag{1.5}$$

where V is the volume of the tetrahedron and $\lambda = +1$ or -1 is the curvature of the space $\mathbb{S}^3$ or $\mathbb{H}^3$. The Schlaefli formula establishes a simple relationship among the most important geometric quantities (volume, length and angle) in dimension 3. What Colin de Verdiere found is the following. Take a geometric triangle in $\mathbb{E}^2$ or $\mathbb{S}^2$ or $\mathbb{H}^2$ of inner angles $\theta_1, \theta_2, \theta_3$ so the opposite edges have lengths $r_2 + r_3, r_1 + r_3$ and $r_1 + r_2$. Then there exists a function $C(r_1, r_2, r_3)$ so that

$$\frac{\partial C}{\partial u_i} = \theta_i$$

where $u_i = \ln r_i$ for $\mathbb{E}^2$, $u_i = \ln\cosh(\frac{r_i}{2})$ for $\mathbb{H}^2$ and $u_i = \ln\coth(\frac{r_i}{2})$ for $\mathbb{S}^2$. Furthermore, he proved that C is strictly concave in (u_1, u_2, u_3) for $\mathbb{H}^2$ and concave for $\mathbb{E}^2$.

Here is how Colin de Verdiere proved Thurston-Andreev's rigidity theorem for hyperbolic circle packing metrics. Suppose (S, T) is a closed triangled surface with the set of vertices V. A hyperbolic circle packing metric on (S, T) is given by a radius assignment $r \in \mathbb{R}^V_{>0}$. Introduce a new coordinate $u \in \mathbb{R}^V_{<0}$ so that $u(v) = \ln\cosh(\frac{r(v)}{2})$. The map $u = u(r)$ is a diffeomorphism from $\mathbb{R}^V_{>0}$ to $\mathbb{R}^V_{<0}$.

Define a function $W : \mathbb{R}^V_{<0} \to \mathbb{R}$ by

$$W(u) = \sum_{\{v_i, v_j, v_k\} \in F} C(u_i, u_j, u_k)$$

where the sum is over all triangles in T with vertices v_i, v_j, v_k and $u_i = u(v_i)$. By the construction, W is strictly concave in u and $\nabla W = 2\pi(1, \cdots, 1) - k_0$ where k_0 is the discrete curvature of the metric r. Since for a strictly concave function W defined on an open convex set, the gradient ∇W is injective. This shows Thurston's rigidity theorem.

Our work on the subject is highly influenced by [CV1]. We set our goal as to find all possible 2-dimensional analogous of Schlaefli formula. This was achieved by carefully exam the cosine law and its derivative. As a consequence we obtained a complete list of all 2-dimensional Schlaefli type identities. Those convex or concave action functionals are also classified. These convex functionals are the main ingredients that we use to establish Theorems 1.12, 1.13 and others.

We are also highly influenced by the works of [CV1], [Thu], [Ri],[MaR] and [Le]. Beside the work of [CV1] and [Ri], variational principles on triangulated surfaces have also appeared in Braegger [Br], Leibon [Le], Cohen-Kenyon-Propp [CKP], Bobenko-Springborn [BS], Springborn [Spr], Schlenkar [Sch2], [Lu1], [Lu2], [Lu3] and others. The energy functions used in [CV1],[Ri],[Br],[Le],[CKP] are related to the 3-dimensional volume or its Legendre transform. Even in the work of [CV1], Colin de Verdiere's energy was motivated by the 3-dimensional Schlaefli volume formula [CV2]. Very recently in 2004, motivated by the discrete 2-dimensional integrable system, Bobenko and Springborn [BS] discovered a new collection of energies for triangulated surfaces.

We observe that all energy functions used by Colin de Verdiere, Braegger, Rivin, Cohen-Kenyon-Propp, Leibon, Bobenko-Springborn and Cohen-Kenyon-Propp can be constructed using the cosine law and the Legendre transformation. Furthermore, we show that these known energy functions are special cases of continuous families of energy functions derived from the cosine law. We also show that these families are the complete lists of all localized energies one could construct. All rigidity results in this book are applications of those convex or concave energy functions.

2

Spherical Geometry and Cauchy Rigidity Theorem

2.1 Spherical Geometry and Spherical Triangles

Let $\mathbb{S}^n \subset \mathbb{R}^{n+1}$ be the unit sphere $\{x \in \mathbb{R}^{n+1} | x \cdot x = 1\}$with the induced Riemannian metric d from $\mathbb{R}^{n+1}$. The basic ingredients in the spherical geometry $(\mathbb{S}^n, d)$ are the group of isometries, the geodesics and totally geodesic submanifolds. The group of isometries of $(\mathbb{S}^n, d)$ is $O(n+1)$, the orthogonal linear group. Geodesics in $(\mathbb{S}^n, d)$ are the great circles which are the intersections $\mathbb{S}^n \cap P$ of $\mathbb{S}^n$ with 2-dimensional linear subspaces P in $\mathbb{R}^{n+1}$. Totally geodesic submanifolds in $\mathbb{S}^n$ are $\mathbb{S}^n \cap Q$ where Q are linear subspaces of dimension $\geqslant 2$ in $\mathbb{R}^{n+1}$. Given $v \in \mathbb{S}^{n+1}$, we call $\{x \in \mathbb{S}^n | x \cdot v \geqslant 0\}$ a hemi-sphere in $\mathbb{S}^n$ with pole v where $u \cdot v$ is the standard inner product in $\mathbb{R}^m$. A spherical triangle in $\mathbb{S}^2$ is the intersection of three hemi-spheres whose poles are linear independent.

The area of a spherical triangle was found by Girard in 1625. It states

Theorem 2.1 (Girard). *If the inner angles of a spherical triangle are $\alpha_1, \alpha_2, \alpha_3$, then its area is $\alpha_1 + \alpha_2 + \alpha_3 - \pi$.*

Proof. Note that the area of $\mathbb{S}^2$ is 4π and the area of the intersection of two hemi-spheres is 2θ if their intersection angle is θ. Now let the spherical triangle be A, let B, C, D be the spherical triangles each sharing exactly one edge with A and bounded by the same geodesics bounding A. See Fig. 2.1. For simplicity, we also use A, B, C to denote the area of triangles A, B, C. Then due to the fact that $A \cup B$ is the intersection of two hemi-spheres intersecting at angle α_2, we obtain

$$A + B = 2\alpha_2,$$

similarly,

$$A + C = 2\alpha_1,$$
$$A + D = 2\alpha_3.$$

Finally $A \cup (-B) \cup C \cup D$ is a hemi-sphere where $-B$ is the triangle antipodal to B. It follows that

$$A + B + C + D = 2\pi.$$

By taking the sum of the first three equations and subtracting the last one, we obtain, $A = \alpha_1 + \alpha_2 + \alpha_3 - \pi$. □

Girard's formula was generalized by Poincaré to any even-dimensional spherical simplex.

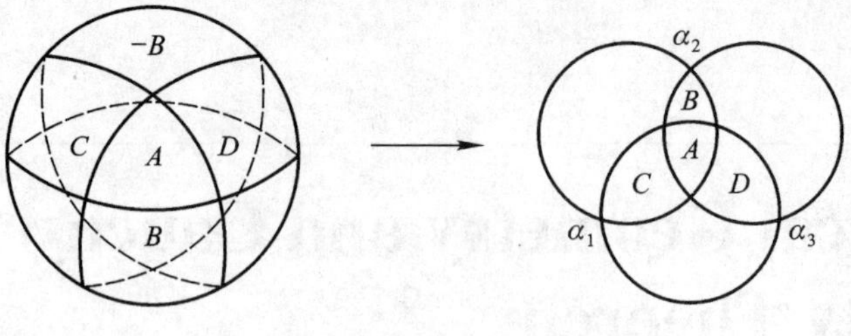

Fig. 2.1. Spherical Triangles

Corollary 2.2. *The area of a sphere polygon P of inner angles* $\alpha_1, \cdots, \alpha_n$ *is* $\sum_{i=1}^{n} \alpha_i - (n-2)\pi$. *See Fig. 2.2.*

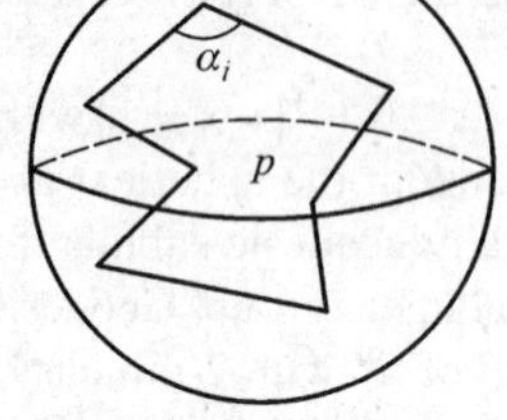

Fig. 2.2.

2.2 The Cosine law and the Spherical Dual

Given a spherical triangle of inner angles $\theta_1, \theta_2, \theta_3$ and edge lengths l_1, l_2, l_3 so that l_i-th edge is opposite to the θ_i-th angle, the cosine law states

$$\cos\theta_1 = \frac{\cos l_1 - \cos l_2 \cos l_3}{\sin l_2 \sin l_3}. \tag{2.1}$$

Proof. Let the spherical triangle be ΔABC and O be the origin. Consider the Euclidean tetrahedron $OPAQ$ where P and Q are in the rays $\overrightarrow{OB}$ and $\overrightarrow{OQ}$ respectively so that both angles $\angle OAP$ and $\angle OAQ$ are $\pi/2$. We will use the Euclidean cosine law:

$$\cos\alpha = \frac{y^2 + z^2 - x^2}{2yz},$$

see Fig. 2.3 (b) for Euclidean triangles.

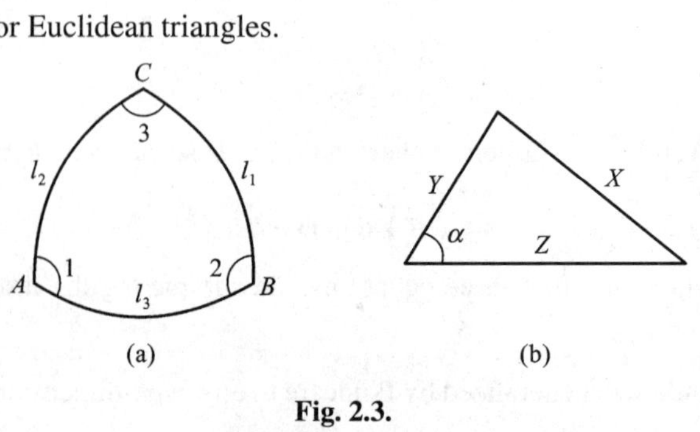

Fig. 2.3.

From Fig. 2.4, we can easily calculate the lengths $|OA|=1$, $|OP|=\frac{1}{\cos l_3}$, $|OQ|=\frac{1}{\cos l_2}$, $|AP|=\tan l_3$, $|AQ|=\tan l_2$,
Using the $\mathbb{E}^2$ cosine law for $\triangle OPQ$, we obtain

$$PQ^2=\frac{1}{\cos^2 l_2}+\frac{1}{\cos^2 l_3}-\frac{2\cos l_1}{\cos l_2\cos l_3}.$$

Using the $\mathbb{E}^2$ cosine law for $\triangle APQ$, we have

$$\begin{aligned}\cos\theta_1 &= \frac{AP^2+AQ^2-PQ^2}{2(AP)(AQ)}\\ &= \frac{\tan^2 l_2-\frac{1}{\cos^2 l_2}+\tan^2 l_3-\frac{1}{\cos^2 l_3}+\frac{2\cos l_1}{\cos l_2\cos l_3}}{2\tan l_2\tan l_3}\\ &= \frac{(-2)+\frac{2\cos l_1}{\cos l_2\cos l_3}}{\frac{2\sin l_2\sin l_3}{\cos l_2\cos l_3}}\\ &= \frac{\cos l_1-\cos l_2\cos l_3}{\sin l_2\sin l_3}.\end{aligned}$$

□

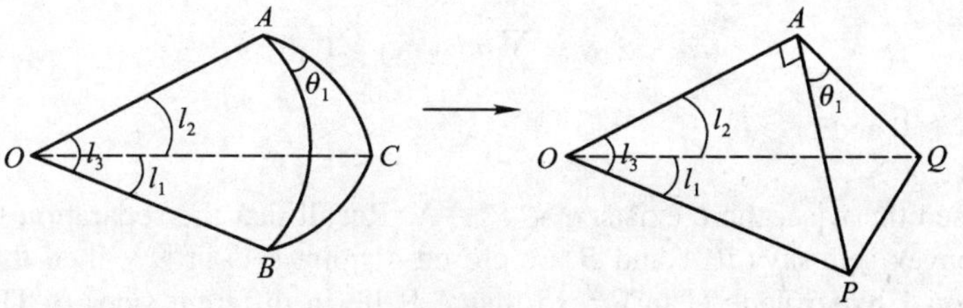

Fig. 2.4.

From the cosine law above, we deduce

$$\cos l_1=\cos l_2\cos l_3+\sin l_2\sin l_3\cos\theta_1.$$

In particular, it shows that $l_1=l_1(l_2,l_3,\theta_1)$ is a function of l_2,l_3 and θ_1. By taking $\frac{\partial}{\partial\theta_1}$ to the above equation, we obtain,

$$\frac{\partial l_1}{\partial\theta_1}=\frac{\sin l_2\sin l_3\sin\theta_1}{\sin l_1}>0.$$

This gives a rigorous proof of the well known fact that if l_2 and l_3 are fixed, then larger θ_1 corresponds to larger l_1.

Another form of the cosine law expresses the length l_i in terms of the angles for spherical triangles,

$$\cos l_i=\frac{\cos\theta_i+\cos\theta_j\cos\theta_k}{\sin\theta_j\sin\theta_k} \tag{2.2}$$

where $\{i,j,k\}=\{1,2,3\}$.

This can be deduced from (2.1) by considering the dual spherical triangle. In fact, the dual triangle has edge lengths $\pi-\theta_i$ and angles $\pi-l_i$ for $i=1,2,3$.

In general, the spherical dual is the following,

Definition 2.3. *A convex polygon $P \subset \mathbb{S}^2$ is the intersection of a finite number of hemispheres, i.e.,*

$$P = \{x \in \mathbb{S}^2 | x \cdot v_i \geqslant 0 \ for\ given\ v_1, \cdots, v_n \in \mathbb{R}^3 - \{0\}\},$$

where $u \cdot v$ is the standard inner product in $\mathbb{R}^3$. The dual P^* of P is defined to be

$$P^* = \{x \in \mathbb{S}^2 | x \cdot y \geqslant 0\ for\ all\ y \in P\}.$$

Given a set of vectors $u_1, \cdots, u_n \in \mathbb{R}^N - \{0\}$, the cone spanned by them is

$$\mathrm{Cone}(u_1, \cdots, u_n) = \{\sum_{i=1}^{n} a_i u_i | a_i \geqslant 0\}.$$

Proposition 2.4. *If $P = \{x \in \mathbb{S}^2 |\ x \cdot v_i \geqslant 0,\ for\ u_1, \cdots, u_n\}$, then*

$$P^* = \mathbb{S}^2 \bigcap \mathrm{Cone}(v_1, \cdots, v_n).$$

Proof. Let $X = \mathbb{S}^2 \bigcap \mathrm{Cone}(v_1, \cdots, v_n)$
(i) $X \subset P^*$ is trivial.
For $x = \sum_{i=1}^{n} a_i v_i \in X$, where $a_i \geqslant 0$ and any $y \in P$,

$$x \cdot y = \sum a_i (v_i \cdot y) \geqslant 0$$

due to $v_i \cdot y \geqslant 0$ and $a_i \geqslant 0$.
(ii) $P^* \subset X$.
Suppose otherwise, there exists $x_0 \in P^* - X$. Recall that the separation theorem for disjoint convex sets says if A and B are closed disjoint sets in $\mathbb{R}^N$, then there exists a codimension-1 hyperplane Π in $\mathbb{R}^N$ so that A,B lie in different sides of Π. Now take $A = \{x_0\}$ and $B = X$, we find a hyperplane Π separating $\{x_0\}$ and X. Since X is a cone with vertex 0, after a translation of Π, we may assume that $0 \in \Pi$, i.e, $\Pi = \{x \in \mathbb{R}^N | x \cdot h = 0\}$ for some $h \in \mathbb{R}^N - \{0\}$. This shows that $x_0 \cdot h < 0$ and $x \cdot h \geqslant 0$ for all $x \in \mathrm{Cone}(v_1, \cdots, v_n)$. Equivalently, we have $v_i \cdot h \geqslant 0$ for all $i = 1, \cdots, n$ and $x_0 \cdot h < 0$. This means $h \in P$ and $x_0 \cdot h < 0$. By definition, it shows that $x_0 \notin P^*$. This contradicts the assumption that $x_0 \in P^* - X$. □

We leave it as a homework to show,

Corollary 2.5. $(P^*)^* = P$.

For example, if $v_1, v_2, v_3 \in \mathbb{S}^2$ are linear independent, then $\triangle = \mathbb{S}^2 \bigcap \mathrm{Cone}(v_1, v_2, v_3)$ is a spherical triangle with vertices v_1, v_2, v_3. Then its dual triangle $\triangle^*$ is $\{x \in \mathbb{S}^2 | x \cdot v_i \geqslant 0\ for\ i = 1, 2, 3\}$. It shows that the vertex v_i of $\triangle$ is the pole of the hemisphere bounding $\triangle^*$. In particular, the spherical distance $d(v_1, v_2)$ is $\pi - \theta_3$ where θ_3 is the inner angle of $\triangle^*$ bounded by $\{x | x \cdot v_1 \geqslant 0\}$ and $\{x | x \cdot v_2 \geqslant 0\}$. This justifies the assertion that lengths and angles of $\triangle^*$ are $\pi - \theta_i$ and $\pi - l_i$ where l_i and θ_i are the lengths and angles of $\triangle$.

2.3 The Cauchy Rigidity Theorem

In this section, we give a proof of the famous Cauchy rigidity theorem. The proof follows closely the one appeared in the book *Polyhedra* by Peter Richard Cromwell[Cro].

Theorem 2.6 (Cauchy Rigidity Theorem). *Suppose P and Q are two compact convex polytopes in $\mathbb{R}^3$ such that there exists an isometry $f: \partial P \to \partial Q$ preserving vertices, edges and faces. Then there exists an isometry F of $\mathbb{E}^3$ so that $F(P) = Q$.*

Proof. It suffices to show that at each edge e of P, the dihedral angle of P at e is equal to the dihedral angle of Q at $f(e)$. In particular, it implies that Cauchy's theorem holds for those P so that each vertex has degree 3. Indeed, in this case, the dihedral angles at edges are determined by the angles in the faces using the cosine law applied to the vertex spherical triangles. □

The strategy of the proof in general goes as follows. Suppose P is not isometric to Q in $\mathbb{E}^3$. We assign each edge e of P the number $1, -1$, or, 0 according to the dihedral angle of P at e is larger, or smaller or equal to that of Q at $f(e)$. Let G be the graph consisting of vertices and edges in P. We consider G as a graph embedded in $\mathbb{S}^2$ (after a homeomorphism from ∂P to $\mathbb{S}^2$).

Cauchy observed two properties of the assignment.

Geometric lemma. *At each vertex v of P, unless all edges ending at v are assigned* 0, *the total change of signs of all edges at v, counted cyclically, is at least 4.*

Topological lemma. *For any 3-connected graph G in $\mathbb{S}^2$, there is no way to assign $+1, -1$, and* 0 *(not all 0) to edges in G so that the conclusion of the geometric lemma holds.*

As a convention, if A, B are two points in $\mathbb{S}^2$, the spherical distance is denoted by $|AB|$.
The proof of the geometric lemma is based on

Cauchy's Arm lemma. *Let $A_1, \cdots, A_n$ and $B_1, \cdots, B_n$ be consecutive vertices of two convex polygons in $\mathbb{E}^2$ or $\mathbb{S}^2$, so that*
(1) $|A_iA_{i+1}| = |B_iB_{i+1}|$, $i = 1, \cdots, n-1$
(2) $\angle A_i \leqslant \angle B_i$, $i = 2, \cdots, n-1$, and not all are equal,
Then $|A_1A_n| < |B_1B_n|$.

Remark. *Cauchy's original proof contains a mistake in that he did not assume that $A_1, \cdots, A_n$ are vertices of a convex polygon. Steinitz found the mistake and fixed it.*

Proof. We use induction on n. If $n = 3$, then $|A_1A_3| < |B_1B_3|$ follows from the fact that $\dfrac{\partial l_1(l_2, l_3, \theta_1)}{\partial \theta_1} > 0$ in the cosine law. Let us assume that $n \geqslant 4$ and that the result has been proved for $n-1$.

There are three cases to be considered.

Case 1. There exists $i \in \{2, \cdots, n-2\}$ so that $\angle A_i = \angle B_i$. Then the result follows by removing the i-th vertex A_i, i.e., we consider $\{A_1, \cdots, A_{i-1}, A_{i+1}, \cdots, A_n\}$ and $\{B_1, \cdots, B_{i-1}, B_{i+1}, \cdots, B_n\}$. By the assumption, these pairs of $n-1$ points satisfy the induction hypothesis. Thus, it follows that $|A_1A_n| < |B_1B_n|$.

Case 2. Suppose $\angle A_i < \angle B_i$, for all $i \in \{2, \cdots, n-1\}$. After composing with an isometry of $\mathbb{S}^2$, we may assume that $A_{n-2} = B_{n-2}$ and $A_{n-1} = B_{n-1}$. Let θ be an angle so that $\angle A_{n-1} < \theta < \angle B_{n-1}$. Let A_θ be point in $\mathbb{S}^2$ so that $|A_\theta A_{n-1}| = |A_nA_{n-1}|$ and $\angle A_{n-2}A_{n-1}A_\theta = \theta$.

condition (*): *For all θ with $\angle A_{n-1} < \theta < \angle B_{n-1}$, $A_1, \cdots, A_{n-1}, A_\theta$ form vertices of a convex polytope.*

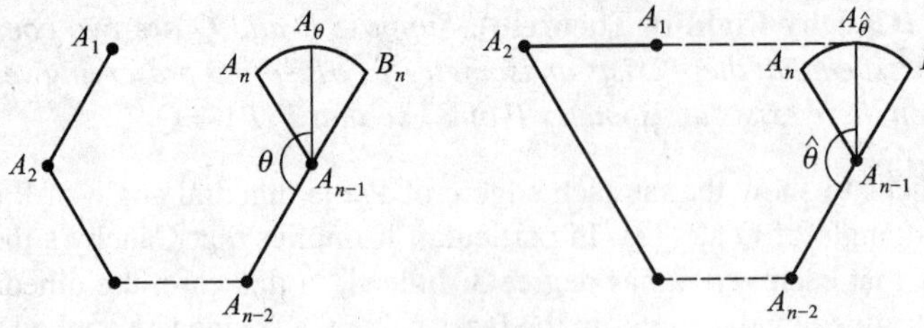

Fig. 2.5.

If condition (*) holds, then by Case 1 applied to $\{A_1,\cdots,A_{n-1},A_n\}$ and $\{A_1,\cdots,A_{n-1},A_\theta\}$, we obtain $|A_1A_n| < |A_1A_\theta|$. Let θ approach $\angle B_{n-1}$. Then A_θ converges to B_n. Thus $|A_1A_n| < |A_1B_n|$.

Case 3. $\angle A_i < \angle B_i$ for all i but condition (*) fails. Let $\widehat{\theta}$ be the smallest angle so that $\angle A_{n-1} < \widehat{\theta} < \angle B_{n-1}$ for which $A_1,\cdots,A_{n-1},A_{\widehat{\theta}}$ are the vertices of an $(n-1)$-side polygon. The minimality of $\widehat{\theta}$ implies that A_1,A_2 and $A_{\widehat{\theta}}$ must be in a geodesic. Now by Case 1 for $\{A_1,\cdots,A_{n-1},A_n\}$ and $\{A_1,\cdots,A_{n-1},A_{\widehat{\theta}}\}$, we have $|A_1A_n| \leqslant |A_1A_{\widehat{\theta}}|$. But $|A_1A_{\widehat{\theta}}| = |A_2A_{\widehat{\theta}}| - |A_1A_2|$. By the induction hypothesis $|A_2A_{\widehat{\theta}}| \leqslant |B_2B_n|$. It follows, by triangle inequality that

$$|A_1A_n| \leqslant |A_2A_{\widehat{\theta}}| - |A_1A_2| \leqslant |B_2B_n| - |B_1B_2| < |B_1B_n|.$$

□

Corollary 2.7. *Suppose X and Y are two convex polygons in $\mathbb{S}^2$ with vertices $A_1,\cdots,A_n$ and $B_1,\cdots,B_n$ respectively so that $|A_iA_{i+1}| = |B_iB_{i+1}|$ for all indices (modulo n). Assign vertex A_i the number $+1,-1$ or 0 according to $\angle A_i > \angle B_i$, $\angle A_i < \angle B_i$ or $\angle A_i = \angle B_i$. If X, Y are not isometric, then the total sign changes along ∂X is least 4.*

Proof. If otherwise that there are only two changes of signs, then one of them is $+1$, and the other -1. Joint two vertices A_i to A_j by a geodesic so that the signs at A_i,A_j are 0 and the two vertices with $+1$ and -1 assigned are in different components of $\partial X - \{A_i,A_j\}$. Then by Cauchy's Arm lemma applied to the two components of $\partial X - \{A_i,A_j\}$, we obtain $|A_iA_j| > |B_iB_j|$ and $|A_iA_j| < |B_iB_j|$. This is a contradiction. □

Now the Geometric lemma follows from the corollary by taking the dual spherical polygons.

A proof of the Topological lemma goes as follows.

Let V,E,F be the sets of all vertices, edges, and faces of $G \subset \mathbb{S}^2$. The Euler characteristic satisfies $|V| - |E| + |F| = 2$ where $|V|$ denotes the cardinality of V.

Case 1. Assume all edges are assigned ± 1. Let F_i be the number of those faces which are i-sided polygons. Then $|F| = \sum\limits_{i\geqslant 3} F_i = \sum\limits_{i\geqslant 2}(F_{2i}+F_{2i+1})$, and $|E| = \frac{1}{2}\Sigma_{i=3} iF_i = \frac{1}{2}\Sigma_{i\geqslant 1}(2iF_{2i}+(2i+1)F_{2i+1})$. If the total sum of all changes counted around all variables is C, by comparing with the sign changes around faces, then we have

$$4V \leqslant C \leqslant \sum_{i\geqslant 2} 2iF_{2i} + \sum_{i\geqslant 1} 2iF_{2i+1}$$

$$\leqslant (2\sum 2iF_{2i} + 2\sum(2i+1)F_{2i+1} - \sum 4F_{2i} - \sum 4F_{2i+1}) = 4E - 4F.$$

the last inequality holds because of $2i \leqslant 4i-4$ for $i \geqslant 2$ and $2i \leqslant 4i+2-4$ for $i \geqslant 1$.

Thus we obtained $|V|-|E|+|F| \leqslant 0$ with contradicts the Euler characteristic formula, $|V|-|E|+|F|=2$.

Case 2. Some assigned numbers are 0. In this case, the argument is essentially the same by focusing on the subgroup G' of G consisting of those edges with ± 1 assigned numbers. We need to use the following version of Euler characteristic formula, namely if G' is any finite graph with k connected components in $\mathbb{S}^2$, then

$$|V|-|E|+|F|=1+k$$

where V, E, F are the sets of vertices, edges and faces of the graph $G' \subset \mathbb{S}^2$. Details will be left to readers.

Remarks:

1. *The same proof works for convex polytopes in* $\mathbb{S}^3$ *and* $\mathbb{H}^3$.

2. *There is a related open question asking if dihedral angles determine the convex polytope in* $\mathbb{S}^3, \mathbb{H}^3$ *or* $\mathbb{E}^3$. *M.Schlenker [Sch] has constructed a two non-isometric convex spherical prisms which have the same combinatorics and the same corresponding dihedral angles. For hyperbolic convex polytopes so that each vertex is adjacent to three edges, C. Hodgson and I. Rivin [HR] proved that dihedral angles determine the convex polytope in* $\mathbb{H}^3$. *J.Stoker [Sto] conjectured that if P,Q are two compact convex polytopes in* $\mathbb{E}^3$ *so that (1) all faces of P and Q are triangles and (2) there exists a homeomorphism h from* ∂P *to* ∂Q *preserving edges and faces and dihedral angles, then P and Q differ by isometry and scaling.*

3

A Brief Introduction to Hyperbolic Geometry

3.1 The Hyperboloid Model of the Hyperbolic Geometry

Recall that the unit sphere $\mathbb{S}^n = \{ x \in \mathbb{E}^{n+1} | (x,x) = 1\}$ is defined by the standard inner product (,) in the Euclidean space $\mathbb{E}^{n+1}$. The hyperbolic n-dimensional space is defined in the similar way using the *Minkowski Space* $\mathbb{E}^{n,1}$ which is $\mathbb{R}^{n+1}$ with inner product $\langle , \rangle$ where

$$\langle x,y \rangle = \sum_{i=1}^{n} x_i y_i - x_{n+1} y_{n+1}$$

with $x = (x_1, \cdots, x_{n+1}), y = (y_1, \cdots, y_{n+1})$. By definition, the n-dimensional (hyperboloid model of) *hyperbolic space* $\mathbb{H}^n$ is $\{x \in \mathbb{E}^{n,1} \mid \langle x,x \rangle = -1,\ x_{n+1} > 0\}$; the *de Sitter space* $\mathbb{S}_1^n$ is defined to be $\{x \in \mathbb{E}^{n,1} \mid \langle x,x \rangle = 1\}$; and the *light Cone* is defined to be $\{x \in \mathbb{E}^{n,1} \mid \langle x,x \rangle = 0\}$.

Example 3.1. *If n = 2, and* $X = (x,y,z)$, *then* $\langle X,X \rangle = x^2 + y^2 - z^2$. *The hyperbolic plane* $\mathbb{H}^2$ *and the de Sitter space are shown in Fig. 3.1.*

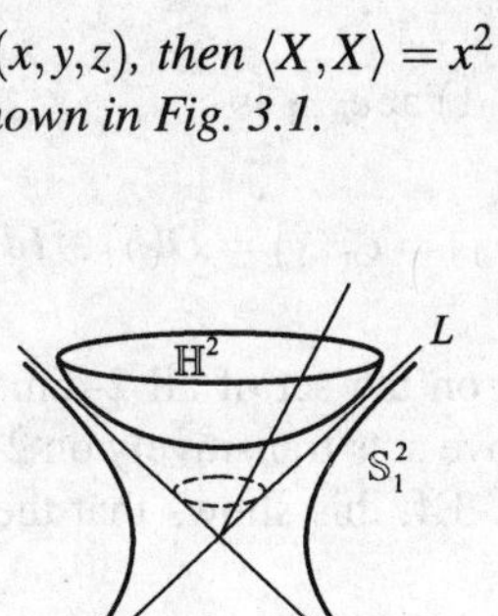

Fig. 3.1.

Lemma 3.2. *The restriction of the bilinear form* $\langle , \rangle$ *to the tangent space* $T_p\mathbb{H}^n$ *is positive definite.*

Proof. Since $\langle p,p \rangle = -1$, the tangent space at p, $T_p\mathbb{H}^2 = \{x \in \mathbb{R}^{n+1} | \langle x,p \rangle = 0\}$. Now suppose otherwise that there exists $x \in T_p\mathbb{H}^2$, such that $\langle x,x \rangle < 0$. By the choice of x and p, we have

(1) $\langle x,x \rangle < 0$, i.e., $\sum_1^n x_i^2 < x_{n+1}^2$,

(2) $\langle p,p\rangle = -1$, i.e., $\sum_1^n p_i^2 + 1 = p_{n+1}^2$, and

(3) $\langle x,p\rangle = 0$, i.e., $\sum_1^n x_i p_i = x_{n+1} p_{n+1}$.

Therefore, by the Cauchy inequality,

$$\left(\sum_1^n x_i^2\right)\left(1+\sum_1^n p_i^2\right) \leqslant x_{n+1}^2 p_{n+1}^2 = \left(\sum_1^n x_i p_i\right)^2 \leqslant \left(\sum_1^n x_i^2\right)\left(\sum_1^n p_i^2\right).$$

This is impossible since $x \neq 0$. □

By this lemma, the restriction $\langle,\rangle|$ produces a Riemannian metric d on $\mathbb{H}^n$. This Riemannian metric d is called *the hyperbolic metric* on $\mathbb{H}^n$.

Definition 3.3. *Let $O(m,1)$ be the subgroup of $GL(m+1,\mathbb{R})$ preserving $\langle,\rangle$ in $\mathbb{E}^{m,1}$. Let $O_+(m,1)$ be the subgroup of $O(m,1)$ preserving $\mathbb{H}^n$. Let $O(m)$ be the orthogonal group of $\mathbb{R}^m$ preserving the standard inner product.*

Lemma 3.4. (a) *The group $O_+(n,1)$ acts on $\mathbb{H}^n$ preserving the Riemannian metric $\langle,\rangle|$.*
(b) *The $O_+(n,1)$-action is transitive on $\mathbb{H}^n$.*

Proof. By the definition of $O_+(n,1)$, we see that it acts isometrically on $\mathbb{H}^n$. To see part (b), let $e_{n+1} = (0,\cdots,0,1) \in \mathbb{H}^n$, and $e_i = (0,\cdots,0,1,0,\cdots,0)$ be the standard basis for $\mathbb{R}^{n+1}$. For any $p \in \mathbb{H}^n$, let $p^\perp = \{x \in \mathbb{R}^{n+1} \mid \langle x,p\rangle = 0\} = T_p\mathbb{H}^n$. Then $e_{n+1}^\perp = \mathbb{R}^n \times 0 \subset \mathbb{R}^{n+1}$. By Lemma 3.2, there is an orthogonal basis $v_1,\cdots,v_n$ in $p^\perp$ with respect to $\langle,\rangle|_{p^\perp}$ so that $\langle v_i,v_j\rangle = \delta_{ij} = \langle e_i,e_j\rangle$ for $i,j \leqslant n$. Define a linear map $\phi : \mathbb{R}^{n+1} \to \mathbb{R}^{n+1}$, by $\phi(v_i) = e_i$, $i = 1,2,\cdots,n$ and $\phi(p) = e_{n+1}$, then $\phi \in O_+(n,1)$. By the construction $\phi(p) = e_{n+1}$, thus (b) follows. □

The isotropy group of $O_+(n,1)$ at e_{n+1} is

$$\{g \in O_+(n,1) \mid g(e_{n+1}) = e_{n+1}\} = O(n) \oplus Id = \{[\begin{smallmatrix} A & 0 \\ 0 & 1 \end{smallmatrix}] | A \in O(n)\}.$$

Since $O(n)$ acts transitively on the set of all 2-dimensional subspaces of $\mathbb{E}^n$, It follows that the isotropy group above acts transitively on 2-dimensional linear subspaces of $T_{e_{n+1}}\mathbb{H}^n$. Together with Lemma 3.4, this shows that the hyperbolic space $(\mathbb{H}^n,d)$ has a constant sectional curvature.

Definition 3.5 (Totally geodesic submanifolds). *If (M,g) is a Riemannian manifold and $S \subset M$ is a k-dimensional smooth submanifold, we say S is totally geodesic in (M,g) if for all geodesics γ in M tangent to S at one point, $\gamma \subset S$.*

Example 3.6. 1*-dimensional totally geodesic submanifolds are the geodesics.*

Example 3.7. *Affine planes $\subset \mathbb{R}^n$ are totally geodesic in $\mathbb{E}^n$. Furthermore, these are all totally geodesic submanifolds in $\mathbb{R}^n$.*

Lemma 3.8. *If Φ is an isometry of (M,g) and $S = \{x \mid \Phi(x) = x\}$, then S is totally geodesic.*

Proof. Say $\gamma(t)$ as above is a geodesic tangent to S at point p. Then $\Phi \circ \gamma(t)$ is another geodesic through p and tangent to γ at p. Since geodesic through a point with a given tangent vector is unique, it follows that $\Phi(\gamma(t)) = \gamma(t)$ for all t. Thus $\gamma \subset S$. □

By considering the isometry $F(x_1,x_2,x_3,\cdots,x_{n+1}) = (-x_1,-x_2,\cdots,-x_{n-1},x_n,x_{n+1})$ of $\mathbb{H}^n$ and its fixed points, we find one geodesic $\mathbb{H}^n \cap (0 \times \mathbb{R}^2)$ in $\mathbb{H}^n$. Using Lemma 3.4, we have

Corollary 3.9. *All geodesics in* $\mathbb{H}^n$ *are of the form* $\mathbb{H}^n \cap (2-\dim$ *linear spaces). All totally geodesic submanifolds in* $\mathbb{H}^n$ *are of the form* $\mathbb{H}^n \cap$ *(linear subspaces).*

Lemma 3.10. (a) *If* $u,v \in \mathbb{H}^n$, *the hyperbolic distance* d *between* u,v *satisfies* $\cosh(d) = -\langle u,v\rangle$.
(b) *If* $u,v \in \mathbb{S}^n_1$*(de Sitter Space), then the dihedral angle* θ *between the codimension-1 totally geodesic hyperplanes* $u^\perp = \{x \in \mathbb{H}^n \mid \langle x,u\rangle = 0\}$ *and* $v^\perp$ *satisfies* $\cos\theta = -\langle u,v\rangle$.

Proof. We will prove part (a) only. Part (b) follows by the same argument. Using the transitive action of $O_+(n+1,1)$, we may assume $u = e_{n+1}, v \in (0,\cdots,0,x,y) \in \mathbb{R}^{n-1} \times \mathbb{R}^2$. Thus, we may assume that $n = 1$. In this case, $u = (0,1), v = (\sinh d, \cosh d)$. Let $\gamma(t) = (\sinh t, \cosh t), t \in [0,d]$ be the geodesic. We have $\gamma'(t) = (\cosh t, \sinh t)$, and $\langle \gamma'(t), \gamma'(t)\rangle = \cosh^2 t - \sinh^2 t = 1$. Thus $\gamma(t)$ is arc-length parameterized. It follows that

$$d(u,v) = \int_0^d \sqrt{\langle \gamma'(t), \gamma'(t)\rangle} dt = d.$$

This shows $\cosh(d) = -\langle u,v\rangle$. □

As a consequence, we see that $(\mathbb{H}^n, \langle,\rangle|)$ is a complete Riemannian manifold since each geodesic can be extended to infinity.

Exercises.
(1) Show that the medias of a hyperbolic triangle intersect in one point.
(2) Can you find a regular dodecahedron(or icosahedron) in $\mathbb{H}^3$?

3.2 The Klein Model of $\mathbb{H}^n$

Consider the open unit disc $\mathbb{D}^n \subset \mathbb{R}^n \times 1 \subset \mathbb{E}^{n,1}$. Define a bijective map $\pi : \mathbb{H}^n \to \mathbb{D}^n$, by sending a point x to $\lambda x \in \mathbb{D}^n$ where $\lambda \in \mathbb{R}_{>0}$. The map π is a radius projection sending geodesics to line segments, totally geodesic submanifolds to affine subspaces.

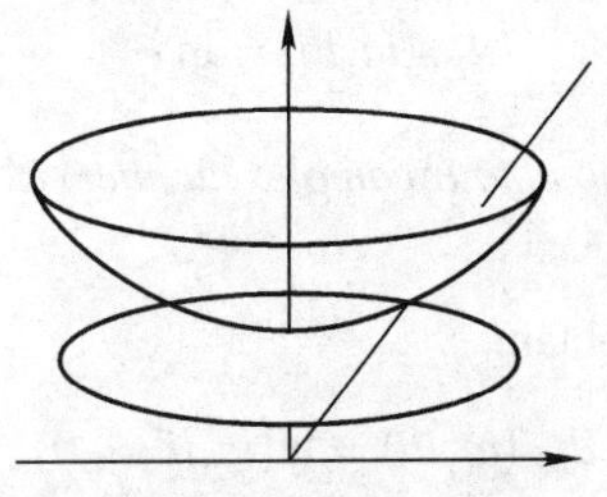

Fig. 3.2.

Make $\mathbb{D}^n$ a Riemannian manifold with the metric so that π is an isometry. Then this Riemannian manifold $\mathbb{D}^n$ is called the *Klein model* of the hyperbolic space. As a consequence, in the Klein model $\mathbb{D}^n$ of $\mathbb{H}^n$, all geodesics are Euclidean line segments inside the open unit disk.

3.3 The Upper Half Space Model of $\mathbb{H}^n$

Let us begin with the 2-dimensional case. The upper half plane is $\mathbb{H}^2 = \{z \in \mathbb{C} \mid Im(z) > 0\}$. Write $z = x + iy$. Define a Riemannian metric on it by $ds^2 = \frac{dx^2+dy^2}{y^2} = \frac{-4dzd\bar{z}}{(z-\bar{z})^2}$. The area form of the metric is $\frac{dxdy}{y^2}$, and the length of a curve γ is $L(\gamma) = \int_a^b \frac{|\gamma'(t)|}{Im(\gamma(t))} dt$. We list all orientation preserving isometries of the metric as follows. First, from the definition of the metric, we see easily that $f(z) = \lambda z$ where $\lambda \in \mathbb{R} > 0$ and $g(z) = z + a$ where $a \in \mathbb{R}$ are isometries. Next we claim that $h(z) = -\frac{1}{z}$ preserves the Riemannian metric. Indeed, let $w = h(z)$. Then,

$$h^*(ds^2) = -\frac{4dwd\overline{w}}{(w-\overline{w})^2} = -\frac{4d(\frac{1}{z})d(\frac{1}{\bar{z}})}{(\frac{1}{z}-\frac{1}{\bar{z}})^2} = -4\frac{\frac{1}{z^2\bar{z}^2}dzd\bar{z}}{\frac{(\bar{z}-z)^2}{z^2\bar{z}^2}} = -4\frac{dzd\bar{z}}{(z-\bar{z})^2}.$$

This verifies that $h(z)$ is an isometry of the hyperbolic metric. It is well known from complex analysis that each Möbius transformation $F(z) = \frac{az+b}{cz+d}, \begin{bmatrix} a & b \\ c & d \end{bmatrix} \in \mathbb{SL}(2,\mathbb{R})$ of the upper half plane is a composition of f's, g's and h's above. It follows that $SL(2,\mathbb{R})$ acts as isometries on $\mathbb{H}^2$. We leave it as an exercise to the reader to verify that all orientations preserving isometries of $\mathbb{H}^2$ are Möbius transformations. (*Hint*: show that $SL(2,\mathbb{R})$ acts transitively on unit tangent vectors.)

Our next task is to find all geodesics in the upper half plane model. Since we know a lot of isometries, it suffices to find one geodesic and then use isometries to find others. Here is a simple calculation.

Example 3.11. *The positive y-axis is a geodesic in* $\mathbb{H}^2$.

Let $\gamma(t)$ be a path from ia to ib, $t \in [0,1]$, in $\mathbb{H}^2$ where $b > a > 0$. Write $\gamma(t) = (x(t), y(t))$, $y(t) > 0$. Then $\gamma'(t) = (x'(t), y'(t))$ and its length is

$$L(\gamma) = \int_0^1 \frac{\sqrt{x'(t)^2 + y'(t)^2}}{y(t)} dt \geqslant \int_0^1 \frac{\sqrt{y'(t)^2}}{y(t)} dt \geqslant \left| \int_0^1 \frac{y'(t)}{y(t)} dt \right| = \ln\frac{b}{a}.$$

Note that equalities hold if and only if $x(t) = 0$ and $y'(t) \geqslant 0$. That is the same as $\gamma(t)$ is monotonic lying in the positive y-axis. This shows that the positive y-axis is a geodesic and the distance between ia and ib is

$$d_{\mathbb{H}^2}(ia, ib) = \ln\frac{b}{a}.$$

Definition 3.12. *The cross ratio of four complex numbers* a,b,c,d *is defined to be* $(a,b,c,d) = \frac{a-c}{a-d} : \frac{b-c}{b-d}$.

Using the cross ratio, we obtain

$$d_{\mathbb{H}^2}(ia, ib) = \ln(ia, ib, \infty, 0).$$

Since Möbius transformations preserve cross ratio and the set of all circles and lines, it follows that the hyperbolic distance between two points $z, w \in \mathbb{H}^2$ is

$$d_{\mathbb{H}^2}(z,w) = \ln(z, w, \widetilde{w}, \widetilde{z}),\ z, w \in \mathbb{H}^2$$

where $\widetilde{z}, \widetilde{w} \in \mathbb{R}$ and $z, w, \widetilde{w}, \widetilde{z}$ are in a circle perpendicular to $\mathbb{R}$. See Fig. 3.3(b) below.

By using the isometrics $\frac{az+b}{cz+d}$, we obtain

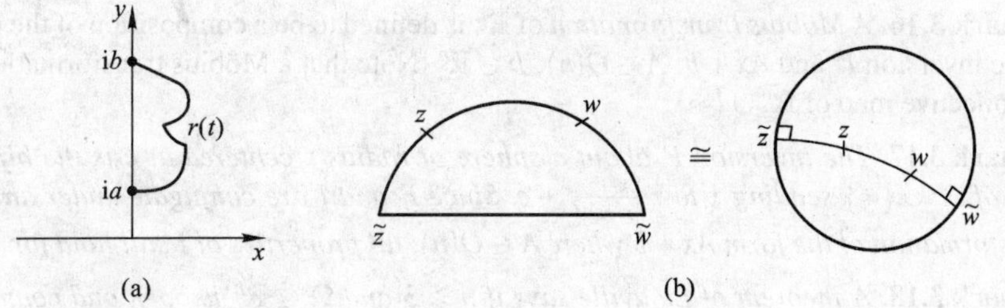

Fig. 3.3.

Corollary 3.13. *All geodesics in $\mathbb{H}^2$ are (portions of) vertical lines or circles perpendicular to the x-axis.*

All of the above computations can be carried out without too much changes to any dimension. The outcome is the upper half space model of of $\mathbb{H}^n$. Here are the definitions. Let the upper half space in $\mathbb{R}^n$ be $\mathbb{H}^n = \{x = (x_1, \cdots, x_{n-1}, t) \in \mathbb{R}^n \mid t > 0\}$. Define a Riemannian metric on it by

$$ds^2 = (dx_1^2 + \cdots + dx_{n-1}^2 + dt^2)/t^2 = (d_E s)^2/t^2.$$

Note that this Riemannian metric is obtained from the Euclidean metric $d_E s^2$ by multiplying by a positive function. Thus the notion of angles in both metrics is the same. This is termed *conformal* in Riemannian geometry and the upper half space model is conformal to the Euclidean geometry. The group of isometries of the Riemannian metric ds can be found as follows. The following maps can be seen from the definition that they preserve the Riemannian metric:

(1) $f(x) = \lambda x$ where λ is a positive real number,
(2) $g(z,t) = (Az, t)$ where $z = (x_1, \cdots, x_{n-1}) \in \mathbb{R}^{n-1}$, $t \in \mathbb{R}_{>0}$ and $A \in O(n-1)$,
(3) $h(z,t) = (z + a, t)$, $a \in \mathbb{R}^{n-1}$.
The inversion I about the unit sphere is an isometry of $\mathbb{H}^n$ given by
(4) $I(x) = \frac{x}{\|x\|^2} = \frac{x}{(x,x)}$.

Lemma 3.14. *The inversion I preserves the Riemannian metric ds^2 in $\mathbb{H}^n$ and in particular it preserves angles in $\mathbb{E}^n$.*

The verification of Lemma 3.14 is calculational. We omit the details.

Among the properties of the inversion, the most interesting one is that the inversion I preserves the set of all (codimension-1) spheres and planes. Here is a proof. The equation of a codimension-1 sphere or plane in $\mathbb{R}^n$ is given by

$$Ax \cdot x + Ba \cdot x + C = 0$$

where $A, B, C \in \mathbb{R}$, $a \in \mathbb{R}^n$, and (x, a) is the standard Euclidean inner product. Now replace x by $\frac{x}{(x,x)}$. The above equation becomes, after a simple substitution,

$$A + Ba \cdot x + Cx \cdot x = 0.$$

It is an equation of a sphere or a plane again.

Remark 3.15. *It can be shown that all isometries of $\mathbb{H}^n$ are compositions of (1),(2),(3) and (4).*

Remark 3.16. *A Möbius transformation* of $\mathbb{R}^n$ is defined to be a composition of the power of the inversion I^k and $Ax+b$, $A \in O(n)$, $b \in \mathbb{R}^n$. Note that a Möbius transformation is a self bijective map of $\mathbb{R}^n \cup \{\infty\}$.

Remark 3.17. *The inversion F about a sphere of radius r centered at c is the bijective map of $\mathbb{R}^n \cup \{\infty\}$ sending x to $r\frac{x-c}{||x-c||^2}+c$. Since F and I are conjugate under an affine transformation of the form $Ax+b$ where $A \in O(n)$, all properties of I still hold for F.*

Remark 3.18. *A theorem of Liouville says if $n \geqslant 3$ and $\Omega \subset \mathbb{R}^n$ is open and connected, then for any angle preserving smooth embedding $f : \Omega \to \mathbb{R}^n$, there exists a Möbius transformation F of $\mathbb{R}^n$ so that $F|_\Omega = f$. It shows the drastic difference between dimension2 and dimension at least 3. For in dimension 2, any injective holomorphic map is a smooth angle preserving embedding.*

By exactly the same argument, we can now find all geodesics in the upper half space model $\mathbb{H}^n$. First using the same argument as in Example 3.11, we prove that the positive t-axis $0 \times \mathbb{R}_{>0}$ is a geodesic in $\mathbb{R}^{n-1} \times \mathbb{R}_{>0}$. Then using isometries and the basic property of Möbius transformations, we conclude that all geodesics in $\mathbb{H}^n$ are either vertical lines or semi-circles perpendicular to $\mathbb{R}^{n-1} \times 0$. All totally geodesic submanifolds in $\mathbb{H}^n$ are hemi-spheres or planes perpendicular to $\mathbb{R}^{n-1} \times 0$.

In the 3-dimensional upper half space model of $\mathbb{H}^3$, a totally geodesic plane in $\mathbb{H}^3$ corresponds to a circle or a line in the complex plane (see Fig. 3.4). Thus a convex hyperbolic polytope in $\mathbb{H}^3$ is given by a circle pattern in $\overline{\mathbb{C}}$. This is the starting point of Thurston's work relating circle packing on surfaces with the 3-dimensional hyperbolic geometry in 1978.

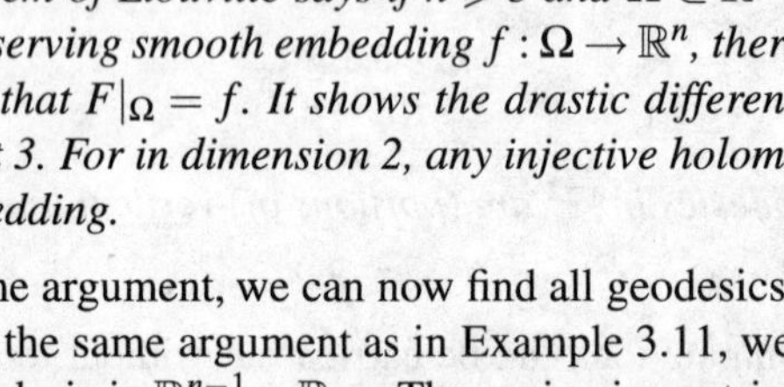

Fig. 3.4.

3.4 The Poincaré Disc Model $\mathbb{B}^n$ of $\mathbb{H}^n$

Let $\Sigma = \{x \in \mathbb{R}^n \mid \| x-(0,\cdots,0,-1) \| = \sqrt{2}\}$ be the sphere of radius $\sqrt{2}$ centered at $c=(0,\cdots,0,-1)$. The sphere Σ intersects both the unit sphere $\partial\mathbb{B}^n = \{x| \, ||x||=1\}$ and the horizontal plane $\mathbb{R}^{n-1} \times 0$ at an angle $\pi/4$. An easily calculation shows that the inversion I_1 about Σ,

$$I_1(x) = \sqrt{2}\frac{x-c}{||x-c||^2}+c$$

sends the upper half space $\mathbb{H}^n$ bijectively onto the open ball $\mathbb{B}^n = \{x| \, ||x|| < 1\}$. Define a Riemannian metric on $\mathbb{B}^n$ by pulling back $(I_1^{-1})^*(ds^2)$. One can show that

$$(I_1^{-1})^*(ds^2) = \frac{4\sum_{i=1}^{n} dx_i^2}{\left(1-\sum_{i=1}^{n} x_i^2\right)^2}.$$

It is highly recommended that the reader carry out the above calculation for $n = 1, 2$. In the case of $n = 2$, one can use the Möbius transformation $-\frac{z-i}{z+i}$ to replace I_1. The pair $(\mathbb{B}^n, (I_1^{-1})^*(ds^2))$ is called the Poincaré disk model of the hyperbolic space. By using the basic properties of the Möbius transformation, we see that all geodesics in $\mathbb{B}^n$ are circular arcs and line segments perpendicular to the unit sphere $\partial\mathbb{B}^n$. All totally geodesic submanifolds are (portions of) spheres and planes perpendicular to $\partial\mathbb{B}^n$.

The advantage of using the Poincaré model is due to its symmetry. For instance, it is obvious from the expression of the Riemannian metric that a hyperbolic ball centered at the origin has to be a Euclidean ball. Using Möbius transformations, one then concludes that all hyperbolic balls are Euclidean balls (even though they may have different centers and radii).

Finally, we should say a word about the equivalences of all these different models. That the upper half space model and the Poincaré models are isometric is due to the construction. So are the Klein model and the hyperboloid model. The isometry between the hyperboloid model and the upper half space model can be established using Cartan's theorem in Riemannian geometry. Cartan's theorem says that any two simply connected and complete Riemannian metrics of constant curvature -1 are isometric. Evidently, both models are simply connected and complete (due to the infinite extendability of the geodesics). Finally, due to the knowledge of the isometry groups, we see that both have constant sectional curvature. It remains to verify that the constant curvature is -1. This can be done by calculating the curvature for $n = 2$ at single point. We leave it to the reader to verify.

3.5 The Hyperbolic Cosine Law and the Gauss-Bonnet Formula

For a hyperbolic triangle of inner angles α, β, γ and edge lengths a, b, c, where angle α faces the edge of length a, the cosine law expresses the relationship between them,

$$\cos\alpha = \frac{\cosh b \cosh c - \cosh a}{\sinh b \sinh c}.$$

Here is a derivation of it. Let the triangle be $\triangle ABC$ (see Fig. 3.5), where $A = 0, B = x, C = ye^{i\alpha}$ in the Poincaré disk model where x, y are positive real numbers. From Fig. 3.5 and the cross ratio formula for hyperbolic distance in §3.5, we can get

$$c = \ln\frac{1+x}{1-x}, \quad \cosh c = \frac{1+x^2}{1-x^2},$$

$$b = \ln\frac{1+y}{1-y}, \quad \cosh b = \frac{1+y^2}{1-y^2},$$

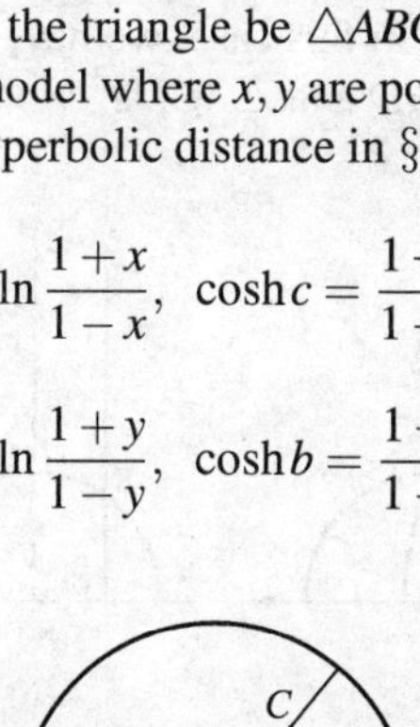

Fig. 3.5. Hyperbolic triangle.

and

$$a = d(x, ye^{i\alpha}) = d(0, \frac{ye^{i\alpha} - x}{xye^{i\alpha} - 1}),$$

where a,b,c are the lengths of the edges. The last equality is due to the isometry $f(z) = \frac{z-x}{xz-1}$ of $\mathbb{D}^2$ so that $f(x) = 0$.

Therefore, we have

$$\begin{aligned}
\cosh a &= \frac{|xye^{i\alpha} - 1|^2 + |ye^{i\alpha} - x|^2}{|xye^{i\alpha} - 1|^2 - |ye^{i\alpha} - x|^2} \\
&= \frac{x^2y^2 + 1 + x^2 + y^2 - xy(e^{i\alpha} + e^{-i\alpha}) - xy(e^{i\alpha} + e^{-i\alpha})}{x^2y^2 + 1 - x^2 - y^2 - xy(e^{i\alpha} + e^{-i\alpha}) + xy(e^{i\alpha} + e^{-i\alpha})} \\
&= \frac{(1+x^2)(1+y^2) - 4xy\cos\alpha}{(1-x^2)(1-y^2)} \\
&= \frac{1+x^2}{1-x^2}\frac{1+y^2}{1-y^2} - \frac{2x}{1-x^2}\frac{2y}{1-y^2}\cos\alpha \\
&= \cosh b \cosh c - \sinh b \sinh c \cdot \cos\alpha.
\end{aligned}$$

□

The Gauss-Bonnet theorem for hyperbolic triangles is the counterpart of Girard theorem in the hyperbolic setting.

Theorem 3.19 (Gauss-Bonnet). *If a hyperbolic triangle in $\mathbb{H}^2$ has inner angles α, β, γ, then the area of the triangle is $\pi - \alpha - \beta - \gamma$.*

Proof. Let the vertices of the triangle be A, B, C (see Fig. 3.6). We may assume that the geodesic AB is a vertical line. Then

$$\triangle ABC = \triangle BCD - \triangle ACD$$

where $D = \infty$. Due to the additivity of area, it suffices to show Gauss-Bonnet formula where $\alpha = 0$.

Next, we may assume further that $\beta = \frac{\pi}{2}$ by dissecting $\triangle ABC$ by the vertical line AD perpendicular to the geodesic BC. In this case, we have either $\triangle ABC = \triangle ABD \cup \triangle ACD$ or $\triangle ABC$ is the difference of $\triangle ABD$ and $\triangle ACD$.

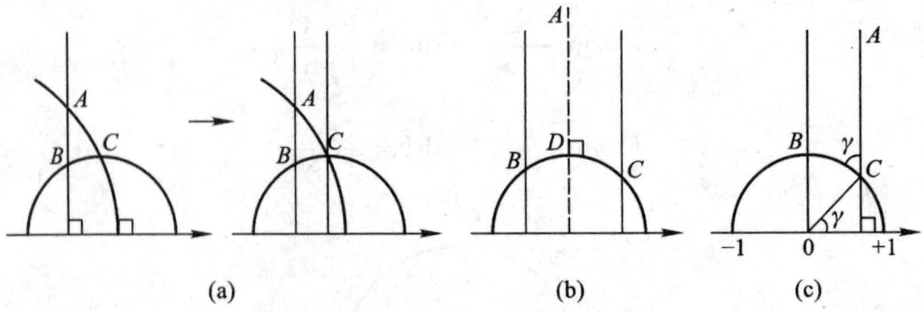

Fig. 3.6.

In the case of $\alpha = 0, \beta = \frac{\pi}{2}$ as shown in figure 3.6(c), we may assume after composing with an isometry that the geodesic AB is the y-axis, the geodesic BC is the unit circle centered at the origin. In this case, the third angle γ is the angle $\angle CO1$, i.e., $C = e^{i\gamma}$. Now the area of

$$\triangle ABC = \{(x,y)|y > \sqrt{1-x^2},\ 0 < x < \cos\gamma\}$$

can be calculated directly as

$$\int_{\triangle ABC} \frac{dxdy}{y^2} = \int_0^{\cos\gamma} \left(\int_{\sqrt{1-x^2}}^{\infty} \frac{dy}{y^2} \right) dx = \int_0^{\cos\gamma} \left(-\frac{1}{y} \right) \Big|_{\sqrt{1-x^2}}^{\infty} dx = \int_0^{\cos\gamma} \frac{1}{\sqrt{1-x^2}} dx$$

$$\overset{x=\cos t}{=} \int_{\frac{\pi}{2}}^{\gamma} \frac{-\sin t}{\sin t} dt = \pi - \left(0 + \frac{\pi}{2} + \gamma\right).$$

This proves the Gauss-Bonnet theorem □

4

The Cosine Law and Polyhedral Surfaces

4.1 Introduction

There are seven cosine laws for triangles and hexagons in $\mathbb{H}^2, \mathbb{S}^2$ and $\mathbb{E}^2$. For simplicity, we take $K^2 = \mathbb{H}^2$ or $\mathbb{S}^2$, or $\mathbb{E}^2$ and let $\lambda = -1, 1$, or 0 be its curvature respectively. Given a triangle $\triangle$ in K^2 with three edges lengths l_1, l_2, l_3 and opposite angles $\theta_1, \theta_2, \theta_3$, the cosine laws express the lengths l_i in terms of $\theta_1, \theta_2, \theta_3$ and vice versa see Fig. 4.1.

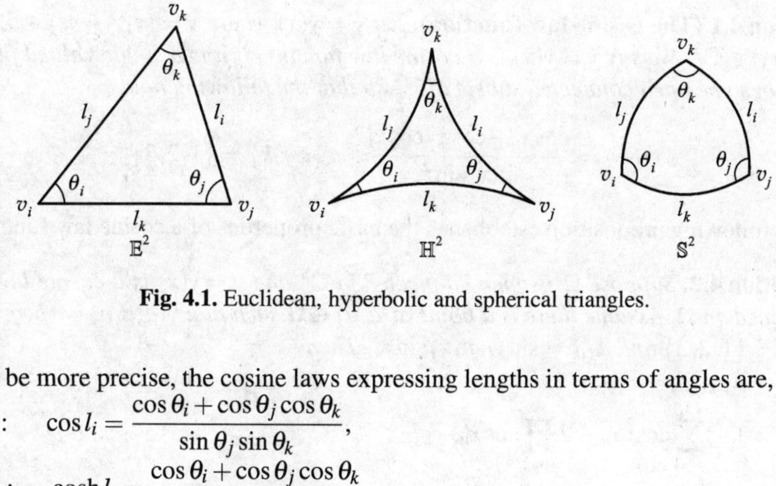

Fig. 4.1. Euclidean, hyperbolic and spherical triangles.

To be more precise, the cosine laws expressing lengths in terms of angles are,

(1) $\mathbb{S}^2: \quad \cos l_i = \dfrac{\cos\theta_i + \cos\theta_j \cos\theta_k}{\sin\theta_j \sin\theta_k}$,

(2) $\mathbb{H}^2: \quad \cosh l_i = \dfrac{\cos\theta_i + \cos\theta_j \cos\theta_k}{\sin\theta_j \sin\theta_k}$,

(3) $\mathbb{E}^2: \quad 1 = \dfrac{\cos\theta_i + \cos\theta_j \cos\theta_k}{\sin\theta_j \sin\theta_k}$,

where $\{i, j, k\} = \{1, 2, 3\}$.

These three cosine laws can be written as:

$$\cos(\sqrt{\lambda}\, l_i) = \frac{\cos\theta_i + \cos\theta_j \cos\theta_k}{\sin\theta_j \sin\theta_k}$$

where $\lambda = 1, -1, 0$ is the curvature of $\mathbb{S}^2$ or $\mathbb{H}^2$ or $\mathbb{E}^2$.

We can interchange the role of edge and angle and get another three cosine laws expressing angles in terms of lengths:

(4) $\mathbb{S}^2: \quad \cos\theta_i = \dfrac{\cos l_i - \cos l_j \cos l_k}{\sin l_j \sin l_k}$,

(5) $\mathbb{H}^2$: $\cos\theta_i = \dfrac{-\cosh l_i + \cosh l_j \cosh l_k}{\sinh l_j \sinh l_k}$,

(6) $\mathbb{E}^2$: $\cos\theta_i = \dfrac{-l_i^2 + l_j^2 + l_k^2}{2 l_j l_k}$.

The last cosine law comes from the right-angled hyperbolic hexagon. Let l_1, l_2, l_3 be three non-pairwise adjacent edges of the hexagon and the opposite edges $\theta_1, \theta_2, \theta_3$.

(7) $\cosh l_i = \dfrac{\cosh\theta_i + \cosh\theta_j \cosh\theta_k}{\sinh\theta_j \sinh\theta_k}$.

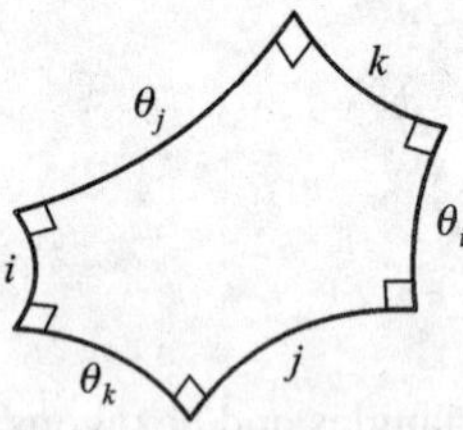

Fig. 4.2.

These seven cosine laws can be put into one form.

Definition 4.1 (The cosine law function). *Let $y = y(x)$ where $y = (y_1, y_2, y_3) \in \mathbb{C}^3$, $x = (x_1, x_2, x_3) \in \mathbb{C}^3$. We say $y = y(x)$ is a cosine law function if it is a single valued function defined in some open connected subset of $\mathbb{C}^3$ so that the following holds,*

$$\cos y_i = \frac{\cos x_i + \cos x_j \cos x_k}{\sin x_j \sin x_k}, \quad \{i,j,k\} = \{1,2,3\} \tag{4.1}$$

The following proposition establishes the basic properties of a cosine law function.

Proposition 4.2. *Suppose Ω is open connected in $\mathbb{C}^3$ and $y = y(x)$ is a cosine law function defined on Ω. Assume there is a point $(a,a,a) \in \Omega$ such that $y(a,a,a) = (b,b,b)$. Let $\{i,j,k\} = \{1,2,3\}$ and $A_{ijk} = \sin y_i \sin x_j \sin x_k$. Then*

(1) $A_{ijk} = A_{123}$,

(2) $A_{ijk}^2 = 1 - \sum_{u=1}^{3} \cos^2 x_u - 2\prod_{u=1}^{3} \cos x_u$,

At point x where $A_{ijk} \neq 0$,

(3) $\dfrac{\partial y_i}{\partial x_i} = \dfrac{\sin x_i}{A_{123}}$,

(4) $\dfrac{\partial y_i}{\partial x_j} = \dfrac{\partial y_i}{\partial x_i} \cos y_k$,

(5) $\cos x_i = \dfrac{\cos y_i - \cos y_j \cos y_k}{\sin y_j \sin y_k}$.

Proof. (2) $A_{ijk}^2 = \sin^2 y_i \sin^2 x_j \sin^2 x_k$

$= (1 - \cos^2 y_i) \sin^2 x_j \sin^2 x_k$

$= \dfrac{\sin^2 x_j \sin^2 x_k - (\cos x_i + \cos x_j \cos x_k)^2}{\sin^2 x_j \sin^2 x_k} \sin^2 x_j \sin^2 x_k$

$= (1 - \cos^2 x_j)(1 - \cos^2 x_k) - \cos^2 x_i - 2\cos x_i \cos x_j \cos x_k - \cos^2 x_j \cos^2 x_k$

$$= 1 - \cos^2 x_j - \cos^2 x_k - \cos^2 x_i - 2\cos x_i \cos x_j \cos x_k.$$

(2) $\Rightarrow$ (1). By (2), $\frac{A_{ijk}}{A_{123}} = \pm 1$, But it takes 1 at (a,a,a) and Ω is connected. It follows that (1) holds.
(3) Take $\frac{\partial}{\partial x_i}$ to equation (4.1). We obtain

$$-\sin y_i \frac{\partial y_i}{\partial x_i} = -\frac{\sin x_i}{\sin x_j \sin x_k}.$$

This is the same as

$$\frac{\partial y_i}{\partial x_i} = \frac{\sin x_i}{\sin y_i \sin x_j \sin x_k} = \frac{\sin x_i}{A_{123}}.$$

(4) Take $\frac{\partial}{\partial x_j}$ to equation (4.1). We obtain

$$\begin{aligned}
-\sin y_i \frac{\partial y_i}{\partial x_j} &= \frac{1}{\sin x_k} \cdot \frac{-\sin^2 x_j \cos x_k - \cos x_j(\cos x_i + \cos x_j \cos x_k)}{\sin^2 x_j} \\
&= \frac{1}{\sin x_k} \cdot \frac{-\sin^2 x_j \cos x_k - \cos x_i \cos x_j - \cos^2 x_j \cos x_k}{\sin^2 x_j} \\
&= \frac{1}{\sin x_k} \cdot \frac{-\cos x_k - \cos x_i \cos x_j}{\sin x_i \sin x_j} \cdot \frac{\sin x_i}{\sin x_j} \\
&= -\frac{\sin x_i}{\sin x_j \sin x_k} \cos y_k,
\end{aligned}$$

It follows

$$\frac{\partial y_i}{\partial x_j} = \frac{\sin x_i}{\sin y_i} \cdot \frac{1}{\sin x_j \sin x_k} \cos y_k = \frac{\partial y_i}{\partial x_i} \cos y_k.$$

(5) $\frac{\cos y_i - \cos y_j \cos y_k}{\sin y_j \sin y_k}$

$$\begin{aligned}
&= \frac{\frac{\cos x_i + \cos x_j \cos x_k}{\sin x_j \sin x_k} - \frac{\cos x_j + \cos x_i \cos x_k}{\sin x_i \sin x_k} \cdot \frac{\cos x_k + \cos x_i \cos x_j}{\sin x_i \sin x_j}}{\sin y_j \sin y_k} \\
&= \frac{(\sin^2 x_i)(\cos x_i + \cos x_j \cos x_k) - (\cos x_j + \cos x_i \cos x_k)(\cos x_k + \cos x_i \cos x_j)}{(\sin y_j \cdot \sin x_i \sin x_k)(\sin y_k \cdot \sin x_i \sin x_j)} \\
&= \frac{(1 - \cos^2 x_i)(\cos x_i + \cos x_j \cos x_k)}{A_{123}^2} \\
&\quad - \frac{(\cos x_j \cos x_k + \cos x_i \cos^2 x_k + \cos x_i \cos^2 x_j + \cos^2 x_i \cos x_j \cos x_k)}{A_{123}^2} \\
&= \cos x_i \cdot \frac{1 - \cos^2 x_i - \cos^2 x_j - \cos^2 x_k - 2\cos x_i \cos x_j \cos x_k}{A_{123}^2} \\
&= \cos x_i \cdot \frac{A_{ijk}^2}{A_{123}^2} \\
&= \cos x_i.
\end{aligned}$$

Note that the identity (5) can be written as

$$\cos(\pi - x_i) = \frac{\cos(\pi - y_i) + \cos(\pi - y_j)\cos(\pi - y_k)}{\sin(\pi - y_j)\sin(\pi - y_k)}. \tag{4.2}$$

Using equations (4.1), (4.2) and (4), we obtain,

$$\frac{\partial x_i}{\partial y_i} = \cos(\pi - x_k)\frac{\partial x_i}{\partial y_i} = -\cos(x_k)\frac{\partial x_i}{\partial y_i} \qquad \square$$

This proposition shows that one can derive identities (4) and (5) from (1) and (2). It also shows that the sine law is a consequence of the cosine law. Since the sine law will be used repeatedly, we state it as

Corollary 4.3. *If $y = y(x)$ is a cosine law function satisfying the condition in Proposition 4.2, then*

$$\frac{\sin(y_i)}{\sin(x_i)} = \frac{\sin(y_j)}{\sin(x_j)}.$$

4.2 Polyhedral Surfaces and Action Functional of Variational Framework

One of the main goals of this book is to study the geometry of polyhedral surfaces. Here is the formal definition,

Definition 4.4 (Polyhedral surfaces). *A Euclidean (or hyperbolic or spherical) polyhedral surface is a triple (S,T,d) where (S,T) is a compact triangulated surface with triangulation T, d is a metric on S so that the restriction of d to each triangle in T is isometric to a Euclidean (or hyperbolic, or spherical) triangle. The metric d is called a polyhedral metric on the triangulated surface (S,T).*

We emphasis that in our approach, the triangulation T is an intrinsic part of the polyhedral surface. Let E be the set of all edges in the triangulation T. A polyhedral metric d on (S,T) is determined by the lengths of the edges, i.e., by the length function $l : E \to \mathbb{R}_{>0}$. In the sequel, we will interchange the use of polyhedral metric d with the edge length function $l : E \to \mathbb{R}_{>0}$. Let V be the set of all vertices in the triangulation T. The classical *discrete curvature* of a polyhedral metric, denoted by k_0, is the function $k_0 : V \to \mathbb{R}$ defined by

$$k_0(v) = 2\pi - \sum_{i=1}^{m} a_i$$

where a_i's are inner angles at the vertex v. The definition of discrete curvature shows that the *basic unit of curvature* in polyhedral surfaces is the inner angle of triangles. The discrete curvature k_0 can be calculated from the metric $l : E \to \mathbb{R}$ by the cosine law for triangles. Thus, *the cosine law should be considered as a metric-curvature relation.* This theme dominates our study of the polyhedral geometry.

A framework for the variational principles on triangulated surfaces is as follows. Fix a triangulated surface (S,T), consider the space X of all polyhedral metrics d's on (S,T). We want to define an action functional on the space X so that the critical point of the functional is the polyhedral metric that we are looking for. The first property of the action functional that we require is the *locality condition*, i.e., the value of the functional at a polyhedral metric is the sum of the action functional of its triangles. This condition is natural from physics point of view. Thus, in order to find all interesting action functional, it suffices to find them for triangles.

Basic Problem 1. (Find all metric-curvature related action functionals) *Given a triangle in $\mathbb{E}^2$ or $\mathbb{H}^2$ or $\mathbb{S}^2$ of edge lengths x_1, x_2, x_3 and opposite inner angles y_1, y_2, y_3, (1) find all smooth functionals $W(x_1, x_2, x_3)$ so that*

$$\frac{\partial W}{\partial f(x_i)} = g(y_i)$$

for some non-constant smooth functionals f, g. This is the same as finding all closed 1-forms of the form $\sum_{i=1}^{3} g(y_i) df(x_i)$.
(2) let $x_i = r_j + r_k$ where $\{i, j, k\} = \{1, 2, 3\}$ or $r_i = \frac{1}{2}(x_j + x_k - x_i)$, find all smooth functionals $W(r_1, r_2, r_3)$ so that

$$\frac{\partial W}{\partial f(r_i)} = g(y_i)$$

for some non-constant functionals f, g. This is the same as finding all closed 1-forms of the form $\sum_{i=1}^{3} g(y_i) df(r_i)$.

This problem, together with its generalization, was solved in [Lu1]. This can be considered as finding all 2-dimensional counterparts of the Schläfli formula. As a consequence, we have a large list of candidates of energy functionals for variational framework on triangulated surfaces. Below is a solution to part (1) of the Basic Problem 1.

Theorem 4.5. *Suppose $y = y(x)$ is a cosine law function. Then all closed 1-forms $\omega = \sum_{i=1}^{3} f(y_i) dg(x_i)$ where f, g are non-constant smooth functions, are up to scaling and complex conjugation, given by*

$$\omega_h = \sum_{i+1}^{3} \left(\int^{y_i} \sin^h t dt \right) d \left(\int^{x_i} \sin^{-h-1} t dt \right) = \sum_{i=1}^{3} \frac{\int^{y_i} \sin^h t dt}{\sin^{h+1} x_i} dx_i$$

where $h \in \mathbb{C}$, i.e., $f(t) = \int^t \sin^h s ds$ and $g(t) = \int^t \sin^{-h-1} s ds$.

Proof. We will only verify that ω_h is closed. The uniqueness is more complicated to prove and is referred to [Lu1]. Write $\omega_h = \sum p_i(x) dx_i$. Then ω_h is closed $\Leftrightarrow \frac{\partial p_i}{\partial x_j} = \frac{\partial p_j}{\partial x_i}$, i.e. $\frac{\partial p_i}{\partial x_j}$ is symmetric in $i, j, i \neq j$, Now the symmetry in our case follows from

$$\begin{aligned}
\frac{\partial}{\partial x_j} \left(\frac{\int^{y_i} \sin^h t dt}{\sin^{h+1} x_i} \right) &= \frac{1}{\sin^{h+1} x_i} \cdot \sin^h y_i \left(\frac{\partial y_i}{\partial x_j} \right) \\
&= \frac{\sin^h y_i}{\sin^{h+1} x_i} \cdot \frac{\sin x_i \cos y_k}{A_{123}} \\
&= \left(\frac{\sin y_i}{\sin x_i} \right)^h \cdot \frac{\cos y_k}{A_{123}}.
\end{aligned}$$

The last expression is symmetric in i, j due to the sine law. □

As a consequence of Theorem 4.5, we have

Corollary 4.6. *For a sphere triangle of length $l = (l_1, l_2, l_3)$ and angle $\theta = (\theta_1, \theta_2, \theta_3)$, then*

$$\omega_h = \sum_{i=1}^{3} \frac{\int^{l_i} \sin^h t dt}{\sin^{h+1} \theta_i} d\theta_i$$

and

$$\omega_h^* = \sum_{i=1}^{3} \frac{\int^{\theta_i} \sin^h t dt}{\sin^{h+1} l_i} dl_i$$

are closed. Furthermore, the integrations $F(u) = \int^u \omega_h$ *and* $G(v) = \int^v \omega_h^*$ *are strictly convex in* u *and* v *respectively, where* $u = (u_1, u_2, u_3)$, $u_i = \int^{\theta_i} \sin^{-h-1} t dt$ *and* $v = (v_1, v_2, v_3)$ *with* $v_i = \int^{l_i} \sin^{-h-1}(t) dt$.

Proof. The closeness is clear form Theorem 4.5 above. It remains to verify that the Hessians of F, G are positive definite. We will prove this for F only. The same argument works for G as well. For F, by definition

$$du_i = \frac{1}{\sin^{h+1}\theta_i} d\theta_i.$$

Thus $\omega_h = \sum_{i=1}^{3} \left(\int^{l_i} \sin^h t dt \right) du_i$. Now,

$$\frac{\partial F}{\partial u_s} = \int^{l_s} \sin^h t dt.$$

Due to $\frac{\partial \theta_r}{\partial u_r} = \sin^{h+1}\theta_r$, we have,

$$\begin{aligned}
\frac{\partial^2 F}{\partial u_r \partial u_s} &= \frac{\partial}{\partial u_r}\left(\int^{l_s} \sin^h t dt \right) = \sin^h l_s \frac{\partial l_s}{\partial u_r} \\
&= \sin^h l_s \frac{\partial l_s}{\partial \theta_r}\frac{\partial \theta_r}{\partial u_r} = \sin^h l_s \frac{\partial l_s}{\partial \theta_r} \cdot \sin^{h+1}\theta_r \\
&= \left(\frac{\sin^h l_s}{\sin^h \theta_s} \right) \cdot (\sin^{h+1}\theta_s \sin^{h+1}\theta_r) \cdot \left(\frac{1}{\sin\theta_s} \frac{\partial l_s}{\partial \theta_r} \right) \\
&= K \cdot C_r C_s \cdot g_{rs}
\end{aligned}$$

where $K = \frac{\sin^h l_i}{\sin^h \theta_i}$ and $C_r = \sin^{h+1}\theta_r$ are some positive functions. Thus, the positivity of the Hessian is equivalent to the positive definiteness of the symmetric matrix $[g_{rs}]_{3\times 3}$.

By Proposition 4.2, we have

$$\begin{aligned}
\frac{\partial l_i}{\partial \theta_i} &= \frac{\sin l_i}{A} = \frac{\sin\theta_i}{\widetilde{A}}, \ A > 0, \widetilde{A} > 0, \\
\frac{\partial l_i}{\partial \theta_j} &= \frac{\partial l_i}{\partial \theta_i} \cos l_k = \frac{\sin\theta_i \cos l_k}{\widetilde{A}},
\end{aligned}$$

and

$$\frac{1}{\sin\theta_i}\frac{\partial l_i}{\partial \theta_j} = \frac{\cos l_k}{\widetilde{A}}.$$

Therefore the matrix

$$[g_{rs}] = \frac{1}{\widetilde{A}} \begin{bmatrix} 1 & \cos l_3 & \cos l_2 \\ \cos l_3 & 1 & \cos l_1 \\ \cos l_2 & \cos l_1 & 1 \end{bmatrix}$$

Let v_1, v_2, v_3 be the three vertices of the sphere triangle. Then we have

$$\cos l_i = v_j \cdot v_k.$$

Thus, the $[g_{rs}]$ is a positive multiplication of the Gram matrix $[(v_r \cdot v_s)]$ which is positive definition. In summary, the Hessian matrix of F is positive definition. Thus the function F is convex in u. □

5

Spherical Polyhedral Surfaces and Legendre Transformation

We establish a rigidity theorem for spherical polyhedral surfaces in this chapter. The Legendre transformation will also be discussed as a tool in variational approach. In the final section, we will discuss some consequences of the Euclidean cosine law.

The framework for our study of polyhedral surfaces is as follows. Fix a closed triangulated surface (S,T). Let E be the set of all edges in the triangulation T. We consider the space of all polyhedral metrics on (S,T). First, an equivalence relation on them is introduced as follows.

Definition 5.1. *Two Euclidean (or Spherical, Hyperbolic) polyhedral metrics on (S,T) are equivalent if there is an isometry between them preserving each edge in T.*

By definition, the equivalence class of a polyhedral metrics on (S,T) is determined by its edge length function $l : E \to \mathbb{R}_{>0}$. Thus, the space of all equivalence classes of polyhedral metrics on (S,T) of $\mathbb{E}^2$, $\mathbb{S}^2$ or $\mathbb{H}^2$ type can be identified with a subset of $\mathbb{R}^E_{>0}$ via edge length functions. In the sequel, we will always identify the space of all equivalence classes with the set of all edge length functions.

5.1 The Space of All Spherical Triangles

The space of all isometry classes of vertex labelled spherical triangles using the length parametrization is

$$S_l(2) = \{(l_1,l_2,l_3) \mid l_i \in (0,\pi),\ l_i + l_j > l_k,\ l_i + l_j + l_k < 2\pi\}.$$

where i,j,k are distinct.
If we use the angle parametrization, the space of all spherical triangles becomes

$$S_\theta(2) = \{(\theta_1,\theta_2,\theta_3) \mid \theta_i \in (0,\pi),\ \theta_i + \theta_j + \theta_k > \pi,\ \theta_i + \theta_j < \theta_k + \pi\}.$$

The cosine law function Π is a diffeomorphism from the open regular tetrahedron $S_l(2)$ to the open regular tetrahedron $S_\theta(2)$. Let $g(x_1,x_2,x_3) = (\pi - x_1, \pi - x_2, \pi - x_3)$ which maps $S_\theta(2)$ to $S_l(2)$. Then the composition $\phi = \Pi \circ g : S_\theta(2) \to S_\theta(2)$ is an involution, i.e., $\phi \circ \phi = id$ (see Fig. 5.1). It is interesting to note that ϕ cannot be extended continuously to the compact closure $\overline{S_\theta(2)}$ to $\overline{S_\theta(2)}$. This is one of the few natural involutions of an open cell which cannot be extended continuously to the compact closure.

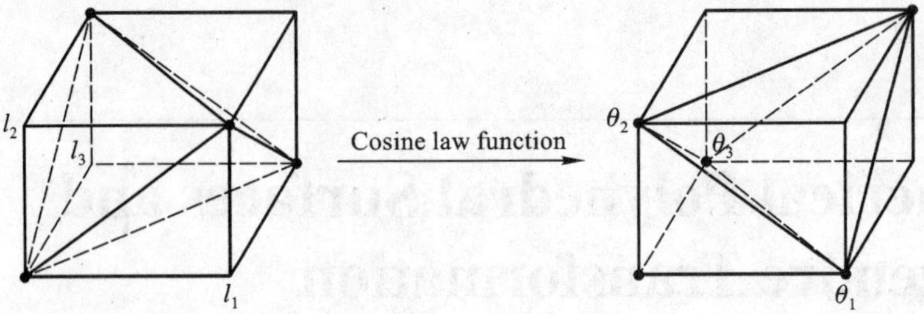

Fig. 5.1.

Exercise 5.2. Verify above statement.

5.2 A Rigidity Theorem for Spherical Polyhedral Surfaces

Recall from Corollary 4.6 that for $l = (l_1, l_2, l_3) \in S_l(2)$,

$$F_0(l) = \int^l \sum_{i=1}^{3} \ln \tan \frac{\theta_i}{2} dl_i$$

is a strictly convex function in $S_l(2)$. Furthermore,

$$\frac{\partial F_0}{\partial l_i} = \ln \tan \frac{\theta_i}{2}.$$

As mentioned before, this is a 2-dimensional analogy of the Schläfli formula. As such, it should be useful for variational framework on triangulated surfaces. Below is just such an application.

Theorem 5.3 (Lu1). *A spherical polyhedral closed surface (S, T, l) is determined up to isometry by its ϕ_{-1} curvature,*

$$\phi_{-1} : \{edges\} \mapsto \mathbb{R}$$

where $\phi_{-1}(e) = \ln \tan \frac{\alpha}{2} + \ln \tan \frac{\beta}{2}$, and α, β are two angles facing the edge e.

The proof to be given below also holds for all ϕ_h-curvature when $h \leqslant -1$. See [Lu1] for details.

Proof. Let $E = \{e_1, \cdots, e_n\}$ be the set of all edges and $T^{(2)}$ be the set of all triangles in the triangulation T. Given a triangle $\sigma \in T^{(2)}$ with edges e_i, e_j, e_k where e_i, e_j, e_k may not be distinct, we use $\{e_i, e_j, e_k\}$ to denote σ. (This notation is ambiguous only when (S, T) is the double of a triangle. In that case, we use $\{e_i, e_j, e_k\}$ to denote any of the two triangles.) By definition a spherical metric l on (S, T) is a map $l : E \to \mathbb{R}_{>0}$ such that if $\{e_i, e_j, e_k\} \in T^{(2)}$, then $(l_i, l_j, l_k) \in S_l(2)$ where $l_i = l(e_i)$. The space of all spherical polyhedral metrics on (S, T), parameterized by the edge lengths, is considered as an open convex polytope, denoted by $\Omega \subset \mathbb{R}^n$. Define an energy $W : \Omega \to \mathbb{R}$ by

$$W(l_1, \cdots, l_n) = \sum_{\{e_i, e_j, e_k\} \in T^{(2)}} F_0(l_i, l_j, l_k).$$

The function W is strictly convex because F_0 is strictly convex. Furthermore,

$$\frac{\partial W}{\partial l_i} = \ln\tan\frac{\alpha}{2} + \ln\tan\frac{\beta}{2} = \phi_{-1}(e_i).$$

where α and β are the angles facing the edge e_i.

Lemma 5.4. *If Ω is open convex in $\mathbb{R}^n$ and $W : \Omega \to \mathbb{R}$ is a smooth strictly convex function with positive definite Hessian matrix, then $\nabla W : \Omega \to \mathbb{R}^n$ is a smooth embedding.*

Proof. If $u \neq v$ in Ω, let $\gamma(t) = (1-t)u + tv \in \Omega$ for all $t \in [0,1]$. Then $g(t) = \omega(\gamma(t)) : [0,1] \to \mathbb{R}$ is a smooth strictly convex function so that

$$\nabla g(t) = \nabla\omega|_{\gamma(t)} \cdot (v-u).$$

Due to strictly convexity, $\nabla g(0) \neq \nabla g(1)$. But $\nabla g(t) = \nabla\omega|_{\gamma(t)} \cdot (v-u)$. It follows that $\nabla W(u) \cdot (v-u) \neq \nabla W(v) \cdot (v-u)$. This shows $\nabla W(u) \neq \nabla W(v)$, i.e., ∇W is injective. On the other hand, the Jacobi matrix of ∇W is the Hessian matrix of W which is non-degenerated. It follows that $\nabla W : \Omega \to \mathbb{R}^n$ is a smooth embedding. □

Remark 5.5. *If Ω is not convex, and* Hess (W) *positive definition, then we can only conclude that ∇W is locally* 1-1 *and is a smooth immersion.*

By using Lemma 5.4, we conclude the proof of Theorem 5.3. □

The argument of the proof of Theorem 5.3 together with Corollary 4.6 and Remark 5.5 immediately imply the following

Theorem 5.6. *For $h \in \mathbb{R}$, a spherical polyhedral surface (S,T,l) is locally determined up to isometry by its ϕ_h-th curvature $\phi_h : \{edges\} \to \mathbb{R}$,*

$$\phi_h(e) = \int_\alpha^{\frac{\pi}{2}} \sin^h t dt + \int_\beta^{\frac{\pi}{2}} \sin^h t dt,$$

where α, β are the angles facing the edge e.

It is proved in [Lu1] that Theorem 5.3 still holds for ϕ_h curvature for all $h \leqslant -1$ or $h \geqslant 0$. For $h \in (0,1)$, we conjecture that the ϕ_h curvature still determines a spherical polyhedral metric up to isometry. Another interesting question relates ϕ_h curvature with hyperbolic polyhedral metrics. Is it true that the ϕ_0 curvature determines a compact hyperbolic tetrahedron?

5.3 The Legendre Transform

We begin with the 1-dimensional case. Suppose $y = f(x)$ is strictly convex in $x \in I$ where I is an interval in $\mathbb{R}^1$. The Legendre transformation $x = g(y)$ of f is a function defined on the image interval $f'(I)$ as follows. For $b \in f'(I)$, consider the line $y = bx$. Then $g(b)$ is the smallest (signed) distance between bx and $f(x)$,

$$g(b) = \max\{bx - f(x)\} = ab - f(a)$$

where $f'(a) = b$ (see Fig. 5.2). For simplicity, let us assume that f is C^2-smooth in the sequel. Then the Legendre transformation $g(y)$ of f is given by

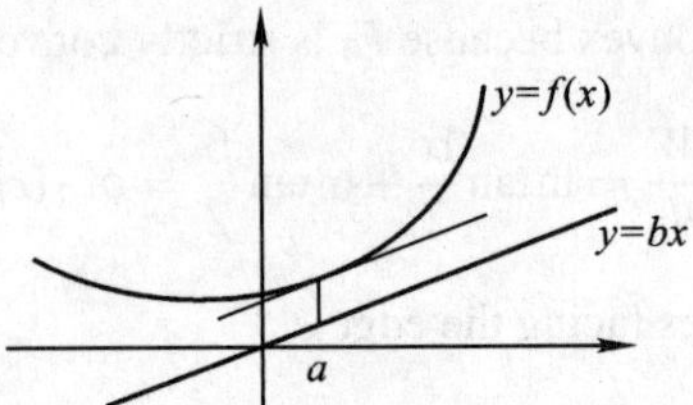

Fig. 5.2.

$$g(y) = xy - f(x) \quad \text{where} \quad f'(x) = y.$$

One important consequence which follows from the definition is the inequality

$$g(y) \geqslant xy - f(x),$$

for $x \in I$ and $y \in f'(I)$. This is the same as

$$f(x) + g(y) \geqslant xy$$

with equality if and only if $y = f'(x)$.

Example 5.7. *Let $f(x) = \frac{x^p}{p}$, $p > 1$, $x > 0$. Take q such that $\frac{1}{p} + \frac{1}{q} = 1$, i.e., $q = \frac{1}{1-\frac{1}{p}} = \frac{p}{p-1}$. Since $f'(x) = x^{p-1}$ and $f''(x) = (p-1)x^{p-2} > 0$, the function f is strictly convex. Its Legendre transformation $x = g(y)$ is given by*

$$g(y) = xy - f(x) = y^{\frac{1}{p-1}} \cdot y - \frac{1}{p}(y^{\frac{1}{p-1}})^p = y^{\frac{p}{p-1}} - \frac{1}{p} y^{\frac{p}{p-1}} = (1 - \frac{1}{p}) \cdot y^q = \frac{y^q}{q}.$$

The basic inequality in this case is the Young's inequality that $\frac{x^p}{p} + \frac{y^q}{q} \geqslant xy$, $\forall x, y > 0$ with equality if and only if $x^p = y^q$.

Example 5.8. *The Legendre transformation of $y = e^x$ is $g(y) = y \ln y - y$.*

Remark 5.9. *Note that the Legendre transformation of $g(y)$ is again strictly convex. Indeed, by definition, $g(y) = xy - f(x)$, $y = f'(x)$. Take derivative $\frac{d}{dx}$ to $g(y) = xy - f(x)$. We obtain, $g'(y)y' = y + xy' - f'(x) = xy'$. Take $\frac{d}{dx}$ to $g'(y) = x$. We have $g''(y)y' = 1$. Since $y' = f''(x)$, therefore $g''(y)f''(x) = 1$. But $f'' > 0$. Thus $g'' > 0$, i.e., g is strictly convex. Furthermore, the above calculation shows, $f(x) = xy - g(y)$, $x = g'(y)$. So the Legendre transformation of g is f.*

Remark 5.10. *The Legendre transformation g of f shows that f' and g' are inverse functions of each other. Indeed, Note that $y = f'(x)$ and $x = g'(y)$, we have $(g')^{-1} = f'$. Furthermore, the Hessians of f and g are reciprocal to each other, i.e., $f''g'' = 1$.*

Remark 5.11. *If $df = f'(x)dx = ydx$, $dg = g'(y)dy = xdy$, then we can write $f = \int^x ydx$, $g = \int^y xdy$.*

All these properties can be generalized to high dimensions. Here is the definition of Legendre transformation in $\mathbb{R}^n$.

Definition 5.12. *Suppose $U,V \in \mathbb{R}^n$ are two domains such that the first de Rahm cohomology group $H^1_{d\mathbb{R}}(U) = 0$ and $y = y(x) : U \to V$ is a diffeomorphism. We assume the Jacobian matrix of $y = y(x)$ is symmetric, i.e.,*

$$\frac{\partial y_i}{\partial x_j} = \frac{\partial y_j}{\partial x_i}.$$

Then, $\omega_U = \sum y_i dx_i$ and $\omega_V = \sum x_i dy_i$ are two closed 1-form in U,V respectively. We call $f(x) = \int^x \omega_U$ and $g(y) = \int^y \omega_V$ Legendre transformations of each other.

Note that $H^1_{d\mathbb{R}}(U) = 0$ implies that f is well defined. Furthermore that $y = y(x)$ is a diffeomorphism shows that $H^1_{d\mathbb{R}}(V) = 0$. In practice, we assume that U,V are simply connected most of the time. In this case $H^1_{d\mathbb{R}}(U) = 0$ is automatic. Note that the Hessian matrix of f, $[\frac{\partial^2 f}{\partial x_r \partial x_s}] = [\frac{\partial y_r}{\partial x_s}]$ is the inverse of the Hessian matrix of g, $[\frac{\partial^2 g}{\partial y_r \partial y_s}] = [\frac{\partial x_r}{\partial y_s}]$.

Example 5.13. *We have two closed 1-forms defined in the space of spherical triangles*

$$\omega_h = \sum_{i=1}^{3} \int^{\theta_i} \sin^h t dt\, d\left(\int^{l_i} \sin^{-h-1} t dt \right),$$

$$\widetilde{\omega}_h = \sum_{i=1}^{3} \int^{l_i} \sin^{-h-1} t dt d\left(\int^{\theta_i} \sin^h t dt \right).$$

One sees that $\int \omega_h$ is the Legendre transform of $\int \widetilde{\omega}_h$.

5.4 The Cosine Law for Euclidean Triangles

Recall that if $y = y(x)$, $x = (x_1, x_2, x_3)$, $y = (y_1, y_2, y_3)$ is a cosine law function

$$\cos y_i = \frac{\cos x_i + \cos x_j \cos x_k}{\sin x_j \sin x_k},$$

{i,j,k}={1,2,3}, then

$$\frac{\partial y_i}{\partial x_j} = \frac{\partial y_i}{\partial x_i} \cos y_k, \quad \frac{\partial x_i}{\partial y_j} = -\frac{\partial x_i}{\partial y_i} \cos x_k.$$

By specializing it to various geometric situations, we obtain more precise results as follows.

Corollary 5.14. *Suppose $l = (l_1, l_2, l_3)$ and $\theta = (\theta_1, \theta_2, \theta_3)$ are lengths and angles of a triangle in $\mathbb{E}^2$, $\mathbb{S}^2$ or $\mathbb{H}^2$, then*

(1) $\dfrac{\partial \theta_i}{\partial l_j} = -\dfrac{\partial \theta_i}{\partial l_i} \cos \theta_k,$

(2) $\dfrac{\partial \theta_i}{\partial l_i} = \begin{cases} \dfrac{l_i}{A}, & \mathbb{E}^2 \\ \dfrac{\sin l_i}{A}, & \mathbb{S}^2 \\ \dfrac{\sinh l_i}{A}, & \mathbb{H}^2 \end{cases}$

where $A > 0$.

Proof. Part (1) is obvious. To see part (2), for $\mathbb{S}^2, \mathbb{H}^2$, we use $y_i = \sqrt{\lambda}\, l_i$, $x_i = \theta_i$ in the formula above. We get

$$\frac{\partial \theta_i}{\sqrt{\lambda}\partial l_j} = -\frac{1}{\sqrt{\lambda}}\frac{\partial \theta_i}{\partial l_i}\cos\theta_k.$$

From this, part (2) follows. □

Euclidean triangles can be considered as degenerated spherical or hyperbolic triangles when they shrink to a point. From this point of view, we can derive the cosine law for Euclidean triangles from that of spherical or hyperbolic ones. Suppose a spherical triangle is given with angles θ_i and lengths l_i. The cosine law for spherical triangle says

$$\cos\theta_i = \frac{\cos l_i - \cos l_j \cos l_k}{\sin l_j \sin l_k}$$

$\{i,j,k\}=\{1,2,3\}$. Now consider a family of spherical triangles of edge lengths $l_i = ta_i$, $a_i > 0$ when $t \to 0$. Then due to the expansion $\cos t = 1 - t^2/2 + o(t^2)$, $\sin t = t + o(t)$,

$$\cos\theta_i = \frac{(1-\frac{a_i^2}{2}t^2) - (1-\frac{a_j^2}{2}t^2)(1-\frac{a_k^2}{2}t^2)}{a_j a_k t^2} = \frac{a_j^2 + a_k^2 - a_i^2}{2a_j a_k}.$$

This is the cosine law for Euclidean triangles of edge lengths a_1, a_2, a_3 and angles $\theta_1, \theta_2, \theta_3$. The Gauss-Bonnet theorem implies that $\theta_1 + \theta_2 + \theta_3 = \pi$.

Let $E_l(2) = \{(l_1, l_2, l_3) \mid l_i + l_j > l_k\}$ be the space of all Euclidean triangles parameterized by edge lengths.

Theorem 5.15. *The following is the complete list of closed 1-forms of the form* $\omega = \sum_{i=1}^{3} f(\theta_i)dg(l_i)$ *for some non-constant smooth functions* f, g *on the space of all Euclidean triangles* $E_l(2) = \{(l_1, l_2, l_3) \mid l_i + l_j > l_k\}$,

$$\omega_h = \sum_{i=1}^{3} \frac{\int^{\theta_i} \sin^h t\, dt}{l_i^{h+1}} dl_i, \quad h \in \mathbb{R}$$

Furthermore, $F_h(u) = \int^u \sum_{i=1}^{3}\left(\int^{\theta_i} \sin^h t\, dt\right) du_i$ *is locally convex in* $u = (u_1, u_2, u_3)$, *where* $u_i = \int^{l_i} t^{-h-1} dt$.

It is proved in [Lu1] that the above is the complete list of closed 1-forms up to scaling. Note that ω_0 was first discovered in [CKP].

Proof. First we verify that ω_h is closed. It is well known that $\omega_h = \sum h_i(l) dl_i$ is closed if and only if $\frac{\partial h_i}{\partial l_j} = \frac{\partial h_j}{\partial l_i}$, i.e., $\frac{\partial h_i}{\partial l_j}$ is symmetric in i, j. Now for $i \neq j$,

$$\begin{aligned}
\frac{\partial}{\partial l_j}\left(\frac{\int^{\theta_i} \sin^h t\, dt}{l_i^{h+1}}\right) &= \frac{1}{l_i^{h+1}} \sin^h \theta_i \frac{\partial \theta_i}{\partial l_j} \\
&= \frac{\sin^h \theta_i}{l_i^{h+1}} \cdot \frac{l_i}{A} \cdot (-\cos\theta_k) \\
&= -\left(\frac{\sin\theta_i}{l_i}\right)^h \cdot \frac{1}{A}\cos\theta_k
\end{aligned}$$

is indeed symmetric in i, j due to the sine law.

Next we check that the Hessian of $F(u)$ is semi-positive definite. We have $du_i = l_i^{-h-1} dl_i$ and $\dfrac{\partial F}{\partial u_r} = \int^{\theta_r} \sin^h t dt$. It follows

$$\frac{\partial^2 F}{\partial u_r \partial u_s} = \sin^h \theta_r \frac{\partial \theta_r}{\partial u_s} = \sin^h \theta_r \frac{\partial \theta_r}{\partial l_s}\frac{\partial l_s}{\partial u_s} = \sin^h \theta_r \frac{\partial \theta_r}{\partial l_s} \cdot l_s^{h+1}$$
$$= \left(\frac{\sin \theta_r}{l_r}\right)^h (l_s \cdot l_r)^{h+1} \left(\frac{1}{l_r}\frac{\partial \theta_r}{\partial l_s}\right).$$

Therefore, it suffices to show the 3×3 matrix $L = [L_{rs}]_{3\times 3} = [\dfrac{1}{l_r}\dfrac{\partial \theta_r}{\partial l_s}]$ is semi-positive definite. We have

$$L_{ii} = \frac{1}{l_i} \cdot \frac{\partial \theta_i}{\partial l_i} = \frac{1}{l_i} \cdot \frac{l_i}{A} = \frac{1}{A}, \ A > 0$$
$$L_{ij} = \frac{1}{l_i} \cdot \frac{\partial \theta_i}{\partial l_j} = \frac{1}{l_i} \cdot \frac{\partial \theta_i}{\partial l_i} \cdot (-\cos\theta_k) = -\frac{\cos\theta_k}{A}, \ i \neq j.$$

Thus,

$$L = \frac{1}{A}\begin{bmatrix} 1 & -\cos\theta_3 & -\cos\theta_2 \\ -\cos\theta_3 & 1 & -\cos\theta_1 \\ -\cos\theta_2 & -\cos\theta_1 & 1 \end{bmatrix}_{3\times 3}.$$

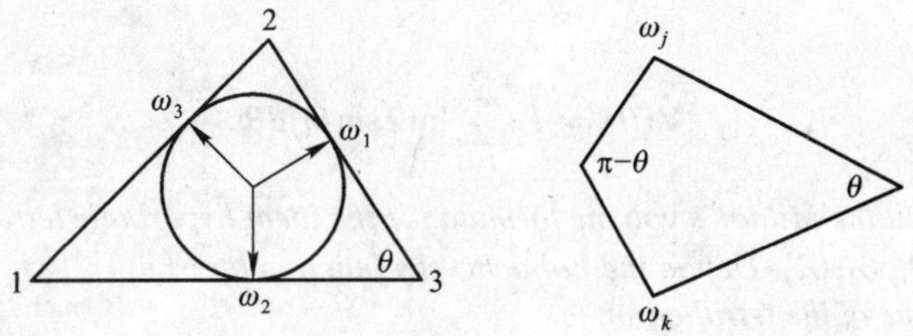

Fig. 5.3.

For a Euclidean triangle with unit outward normal vectors $\omega_1, \omega_2, \omega_3$, we have $\cos\theta_i = -(\omega_j, \omega_k)$, $1 = (\omega_i, \omega_i)$. Thus, $L = \dfrac{1}{A}[(\omega_r, \omega_s)]$ is semi-positive definite. Furthermore, the rank of L is 2. The null space of L is generated by $(\sin\theta_i, \sin\theta_j, \sin\theta_k) = \mu(l_i, l_j, l_k)$. □

Corollary 5.16. *The null space of* $\mathrm{Hess}(F_h)|_u$ *is* $\{ku \mid k \in \mathbb{R}\}$ *if* $h \neq 0$ *and is* $\{k(1,1,1) \mid k \in \mathbb{R}\}$ *if* $h = 0$.

Indeed, if $(v_1, v_2, v_3)\mathrm{Hess}(F_h)(v_1, v_2, v_3)^T = 0$, then

$$(l_1^{h+1} v_1, l_2^{h+1} v_2, l_3^{h+1} v_3)\mathrm{Hess}(F_h)(l_1^{h+1} v_1, l_2^{h+1} v_2, l_1^{h+1} v_3)^T = 0.$$

This implies

$$l_i^{h+1} v_i = \mu \cdot l_i$$

or the same,

$$v_i = l_i^{-h} \cdot \mu.$$

On the other hand, $u_i = \int^{l_i} t^{-h-1} dt = (-\frac{1}{h}) l_i^{-h}$. Therefore, $v_i = \ln\mu$ if $h = 0$, and $v_i = u_i \mu$ if $h \neq 0$.

Example 5.17. *For a Euclidean triangle of edge lengths l_i and angles θ_i, $\omega_0 = \sum_{i=1}^3 \frac{\theta_i}{l_i} dl_i = \sum_{i=1}^3 \theta_i d\ln l_i$ and $\omega_{-1} = \sum_{i=1}^3 \ln\tan\frac{\theta_i}{2} dl_i$ are closed. The function $F_0 = \int \omega_0$ was founded by Cohen, Kenyon, Paulon [CKP] in their work on dimer model in 1999. The function*

$$F_h(u) = \int^u \sum_{i=1}^3 \int^{\theta_i} \sin^h t dt\, du_i, \quad u_i = \int^{l_i} t^{-h-1} dt$$

is not strictly convex on $E_l(2)$. However, with a little care, one can still define the Legendre transformation of F_0. Indeed, it is easy to see that $F_0|_X$ is strictly convex where $X = \{(u_1, u_2, u_3) \mid u_1 + u_2 + u_3 = 0\}$. Note that $u_1 + u_2 + u_3 = 0$ is equivalent to $l_1 l_2 l_3 = 1$. In this case,

$$F_0 = \int \sum_{i=1}^3 \theta_i d\ln l_i.$$

The Legendre transformation of $F_0|_X$ is

$$V(\theta) = \int^\theta \sum_{i=1}^3 \ln l_i d\theta_i$$

defined on $\{(\theta_1, \theta_2, \theta_3) \mid \theta_i > 0,\ \theta_1 + \theta_2 + \theta_3 = \pi\}$. By the sine law, $l_i = 2A\sin\theta_i$, we have

$$\sum_{i=1}^3 \ln l_i d\theta_i = \sum_{i=1}^3 \ln(2\sin\theta_i) d\theta_i + (\ln A \sum_{i=1}^3 d\theta_i) = \sum_{i=1}^3 \ln(2\sin\theta_i) d\theta_i.$$

Thus we obtain

$$V(\theta) = \int^\theta \sum_{i=1}^3 \ln(2\sin\theta_i) d\theta_i.$$

Let us recall the Milnor's volume formula for the ideal hyperbolic tetrahedra of dihedral angles $(\theta_1, \theta_2, \theta_3)$. Define the Lobachevsky function by $\wedge(f) = \int_0^f -\ln(|2\sin u|) du$. Then the volume of the tetrahedron is

$$V(\theta_1, \theta_2, \theta_3) = -\wedge(\theta_1) - \wedge(\theta_2) - \wedge(\theta_3).$$

It follows that the Legendre transformation of $F_0|_X$ is the negative of the volume of the ideal hyperbolic tetrahedra with dihedral angles $\theta_1, \theta_2, \theta_3$. This is the energy functional used by I. Rivin [Ri] in 1995 in his variational principle for Euclidean polyhedral surfaces.

We end this chapter with a question. What is the geometric meaning of $\int^u \omega_h$ or its Legendre transformation?

6
Rigidity of Euclidean Polyhedral Surfaces

We will use the variational principle to establish several rigidity theorems for Euclidean polyhedral surfaces. The energy functions are all derived from the cosine law and the Legendre transformations.

The following notion will be used in this section. Two Euclidean polyhedral metrics d_1, d_2 on a triangulated surface (S,T) are called *homothetic* if there is a constant $k>0$ so that d_1 and kd_2 are equivalent. Thus if the space Y of all Euclidean polyhedral metrics on (S,T) is considered as a subset of $\mathbb{R}^E_{>0}$ via length parameterization, then the space of all homothety classes of all Euclidean polyhedral metrics on (S,T) can be identified with $\{(x_1,\cdots,x_{|E|})\in Y|\sum_{i=1}^{|E|}x_i=1\}$. Recall that for $h\in\mathbb{R}$, the ϕ_h curvature of a polyhedral metric on (S,T) is the function $\phi_h: E\to\mathbb{R}$ so that

$$\phi_h(e)=\int_\alpha^{\frac{\pi}{2}}\sin^h t dt+\int_\beta^{\frac{\pi}{2}}\sin^h t dt$$

where α,β are the angles facing the edge e.

6.1 A Local and a Global Rigidity Theorem

We will prove the following result.

Theorem 6.1 ((Lu1)). *Let (S,T) be a closed triangulated connected surface.*
(a) *For $h\in\mathbb{R}$, a Euclidean polyhedral surface (S,T,l) is locally determined up to homothety by its ϕ_h curvature.*
(b) *If $h\leqslant -1$, then a Euclidean polyhedral metric is determined by its ϕ_h curvature.*

It is proved in [Lu1] that if $h\leqslant -1$, the set of all ϕ_h curvatures of polyhedral metrics on (S,T) forms a complete smooth codimention 1 submanifold in $\mathbb{R}^{\mathbb{E}}$.

Proof. Let $E=\{e_1,\cdots,e_n\}$ be the set of all edges in the triangulation T and $T^{(2)}$ be the set of all triangles in T. A triangle $\sigma\in T^{(2)}$ with edges e_i,e_j,e_k will be denoted by $\{e_i,e_j,e_k\}$. By definition, a Euclidean polyhedral metric on (S,T) is given by its length function $l:E\to\mathbb{R}_{>0}$. We write the metric as $l=(l_1,\cdots,l_n)\in\mathbb{R}^n_{>0}$ where $l_i=l(e_i)$ is the length of the i-th edge. Let $E_l(2)=\{(l_1,l_2,l_3)\mid l_i+l_j>l_k\}$ be the space of all Euclidean triangles parameterized by the edge lengths. The space of all equivalent classes of Euclidean polyhedral metrics on (S,T) is

$$Y = \{l = (l_1, \cdots, l_n) \in \mathbb{R}^n \mid \text{whenever } \{e_i, e_j, e_k\} \in T^{(2)}, \text{ then } (l_i, l_j, l_k) \in E_l(2)\}.$$

Make a change of variable from l to u as follows. If $h \neq 0$, define $u_i = \int^{l_i} t^{-h-1} dt = -\frac{1}{h} l_i^{-h}$; if $h = 0$, define $u_i = \ln l_i$. Let $u = (u_1, \cdots, u_n)$. Note that $\frac{du_i}{dl_i} > 0$ for $l_i > 0$ and $u_i(kl_i) = k^{-h} u_i(l_i)$ for $h \neq 0$, $u_i(kl_i) = u_i(l_i) + \ln k$ for $h = 0$. By the construction, the map $u = u(l)$ is a diffeomorphism. The image of Y under $u = u(l)$ is denoted by

$$\Omega = \{u = (u_1, \cdots, u_n) \in \mathbb{R}^n \mid u = u(l),\ l \in Y\}.$$

By the construction, Ω is another parameter space of the space of all Euclidean polyhedral metrics on (S, T).

Define an energy function $W : \Omega \to \mathbb{R}$ by

$$W(u_1, \cdots, u_n) = \sum_{\{i,j,k\} \in T^{(2)}} F_h(u_i, u_j, u_k)$$

where $F_h(u_1, u_2, u_3) = \int_a^{(u_1,u_2,u_3)} \sum_{i=1}^{3} \left(\int_{\theta_i}^{\frac{\pi}{2}} \sin^h t dt \right) du_i$ and $a = (-1, -1, -1)$ for $h \neq 0$ and $a = 0$ for $h = 0$. By the construction, $\nabla W = (\phi_h(e_1), \cdots, \phi_h(e_n))$ is the ϕ_h curvature and W is locally concave since F is concave by Theorem 5.15.

We need to discuss only two cases to finish the proof of part (a).

Case 1. $h \neq 0$. Then due to $u_i(kl_i) = k^{-h} u_i(l_i)$, we have $\Omega = k\Omega$ for all $k > 0$. (Recall if $Z \subset \mathbb{R}^m$, then $kZ = \{kx | x \in Z\}$.) Let $X = \{u = (u_1, \cdots, u_n) \in \Omega \mid \sum u_i = -1\}$. Each point in X corresponds to a homothety class of polyhedral metric on (S, T).

Lemma 6.2. *The restriction $W| : X \to \mathbb{R}$ is locally strictly convex.*

Assuming this lemma, since X is open in the affine plane $\{u \in \mathbb{R}^n \mid \sum_{i=1}^n u_i = -1\}$, it follows that $\nabla(W)| : X \to \mathbb{R}^{n-1}$ is a local diffeomorphism. But $\nabla(W|) = \text{Proj}(\nabla W|)$ where Proj is the orthogonal projection from $\mathbb{R}^n$ to the affine plane. It follows that $(\nabla W)| : X \to \mathbb{R}^n$ is locally injective. This establishes the local rigidity in the theorem.

Proof (Lemma 6.2). To prove the lemma, it suffices to check that the Hessian Hess $(W|)$ is positive definite. By definition, it is equivalent to show that the null space of the Hessian Hess $(W)|_u$ intersects the tangent space $T_u X$ only at the origin.

Suppose $v = (v_1, \cdots, v_n)$ is a vector in the null space of Hess (W) at the point u, then $v \cdot \text{Hess}(W)_u \cdot v^t = 0$. This is the same as

$$\sum_{\{e_i,e_j,e_k\} \in T^{(2)}} (v_i, v_j, v_k) \cdot \text{Hess}(F_h)|_{(u_i,u_j,u_k)} \cdot (v_i, v_j, v_k)^t = 0.$$

Since F_h is locally convex, $(v_i, v_j, v_k) \cdot \text{Hess}(F_h)|_{(u_i,u_j,u_k)} \cdot (v_i, v_j, v_k)^t \geqslant 0$. It follows that

$$(v_i, v_j, v_k) \cdot \text{Hess}(F_h)|_{(u_i,u_j,u_k)} \cdot (v_i, v_j, v_k)^t = 0.$$

By the Corollary 5.16, there exists $C_{ijk} \in \mathbb{R}$ symmetric in i, j, k so that

$$(v_i, v_j, v_k) = C_{ijk}(u_i, u_j, u_k).$$

This implies that for triangle $\{e_i, e_j, e_k\}$ sharing an edge with triangle $\{e_i, e_l, e_m\}$, we have

$$(v_i, v_l, v_m) = C_{ilm}(u_i, u_l, u_m).$$

By comparing with v_i, we have $v_i = C_{ijk}u_i$ and $v_i = C_{ilm}u_i$. Since $u_i \neq 0$, it follows that $C_{ijk} = C_{ilm}$. Since the surface S is connected, this implies that $C_{ijk} = c$ is a constant independent of the triangles, that is to say

$$v = c \cdot u.$$

This shows $v \notin T_uX$ unless $v = 0$.

Case 2. $h = 0$. In this case, take $X = \{u \in \Omega \mid \sum u_i = 0\}$ which corresponds to the space of all homothety classes of Euclidean polyhedral metrics on (S,T). The same proof goes through as in Case 1 where $v = (v_1, \cdots, v_n)$ satisfies $v_1 = \cdots = v_n$.

To prove part (b) on the global rigidity for $h \leqslant -1$, by Lemma 5.4, it suffices to show that the space $\Omega \subset \mathbb{R}^n$ is convex. Indeed, it suffices to show that the image $\Omega(2)$ of $E_l(2) = \{(l_1, l_2, l_3) \in \mathbb{R}^3 \mid l_i + l_j > l_k\}$ under $u = (u_1, u_2, u_3) = u(l)$ is convex in $\mathbb{R}^3$. Indeed, by definition, the space Ω is the intersection of $\Omega(2) \times \cdots \times \Omega(2)$ with affine spaces.

To establish the convexity of $\Omega(2)$, let us write $u_i = p(l_i)$ where p is a smooth increasing function with $p' > 0$, where $l_i \in \mathbb{R}_{>0}$, $u_i \in \mathbb{R}_{<0}$. Let $g = p^{-1}$ be the inverse of p. Then

$$\begin{aligned}\Omega(2) &= \{(u_1, u_2, u_3) \in \mathbb{R}^3 \mid g(u_i) + g(u_j) > g(u_k)\} \\ &= \{(u_1, u_2, u_3) \in \mathbb{R}^3 \mid p(g(u_i) + g(u_j)) > u_k\}.\end{aligned}$$

Thus, if the function $f(x,y) = p(g(x) + g(y))$ is concave in x, y, then $\Omega(2)$ is the intersection of three convex sets and therefore $\Omega(2)$ is convex.

To check the concavity of f, we need to calculate the Hessian of f. We summarize the calculation in the following lemma which will be used in several places.

Lemma 6.3. *Suppose p is a smooth function so that $p' > 0$ and g is the inverse of p. Let $f(x,y) = p(g(x) + g(y))$ and $u = g(x), v = g(y)$, $w = u + v$. Then*
(a)

$$\frac{\partial^2 f}{\partial x^2} = \frac{p'(w)}{p'(u)^2}(A(w) - A(u))$$

where $A(u) = \dfrac{p''(u)}{p'(u)}$, *and*
(b) *the determinant of the Hessian is*

$$\left(\frac{p'(w)^2}{p'(u)^2 p'(v)^2}\right)(A(u)A(v) - A(w)(A(u) + A(v))).$$

Proof. We have

$$g'(x) = \frac{1}{p'(u)}, \quad g''(x) = -\frac{p''(u)}{p'(u)^3}$$

The second derivative of f is

$$\begin{aligned}
\frac{\partial f}{\partial x} &= p'(w)g'(x)\\
\frac{\partial^2 f}{\partial y\partial x} &= p''(w)g'(x)g'(y) = \frac{p''(w)}{p'(u)p'(v)}\\
\frac{\partial^2 f}{\partial x^2} &= p''(w)g'(x)^2 + p'(w)g''(x)\\
&= \frac{p''(w)}{p'(u)^2} - \frac{p'(w)p''(u)}{p'(u)^3}\\
&= (\frac{p'(w)}{p'(u)^2})(\frac{p''(w)}{p'(w)} - \frac{p''(u)}{p'(u)})\\
&= \frac{p'(w)}{p'(u)^2}(A(w)-A(u)).
\end{aligned}$$

The determinant of the Hessian is

$$\begin{aligned}
&\frac{p'(w)^2}{p'(u)^2p'(v)^2}[(A(w)-A(u))\cdot(A(w)-A(v))] - \frac{p''(w)^2}{p'(u)^2p'(v)^2}\\
&= (\frac{p'(w)^2}{p'(u)^2p'(v)^2})(A(w)^2 + A(u)A(v) - A(w)(A(u)+A(v)) - A(w)^2)\\
&= (\frac{p'(w)^2}{p'(u)^2p'(v)^2})(A(u)A(v) - A(w)(A(u)+A(v))).
\end{aligned}$$

□

Now in our case, $p(x) = -\frac{1}{h}x^{-h}$, $p'(x) = x^{-h-1}$, $p''(x) = (-h-1)x^{-h-2}$. Thus $A(x) = -\dfrac{h+1}{x}$. The determinant of the Hessian of f is

$$\left(\frac{p'(w)^2}{p'(u)^2p'(v)^2}\right)(h+1)^2\left(\frac{1}{u}\cdot\frac{1}{v} - \frac{1}{u+v}\left(\frac{1}{u}+\frac{1}{v}\right)\right) = 0.$$

The second derivative $\dfrac{\partial^2 f}{\partial x^2}$ is

$$q\cdot(-h-1)\left(\frac{1}{w}-\frac{1}{u}\right) = q\cdot(-h-1)\left(\frac{-v}{wu}\right), \quad \text{where} \quad q>0.$$

Note $u,v,w>0$, by definition. It follows that $\dfrac{\partial^2 f}{\partial x^2} \leqslant 0$ is equivalent to $-h-1 \geqslant 0$, i.e., $h \leqslant -1$. This shows that for $h \leqslant -1$, Ω is a convex set. □

6.2 Rivin's Theorem on Global Rigidity of ϕ_0 Curvature

The energy function $F_0(l_1,l_2,l_3) = \int \sum_{i=1}^3 \theta_i d(\ln l_i)$ appeared in the work of Cohen-Kenyon-Propp [CKP] in 1999. The Legendre transformation of it was discovered earlier in 1995 by I. Rivin [Ri] who used it to prove a global rigidity theorem for ϕ_0 curvature. We will follow the proof in [Ri] in this section.

Theorem 6.4 (Rivin). *A Euclidean polyhedral surface (S,T,l) is determined up to scaling by its ϕ_0 curvature which sends an edge e to $\pi-\alpha-\beta$ where angles α,β face the edge e.*

The counterpart of Rivin's theorem for spherical polyhedral metrics was established by Luo in [Lu2]. To be more precise, it is proved in [Lu2] that a spherical polyhedral surface (S,T,l) is determined up to isometry by its ϕ_0 curvature. It is not known if ϕ_0 curvature determines a hyperbolic polyhedra metric.

Proof (Rivin). The proof uses the Legendre transformation of the energy function $F_0|$ restricted to $l_1+l_2+l_3=1$. The calculation at the end of Chapter 5 shows that the Legendre transformation of $-F_0|$ is the volume $V(\theta_1,\theta_2,\theta_3)$ of the hyperbolic ideal tetrahedron with dihedral angles $\{\theta_1,\theta_2,\theta_3,\theta_1,\theta_2,\theta_3\}$, i.e.,

$$V(\theta)=\wedge(\theta_1)+\wedge(\theta_2)+\wedge(\theta_3)$$

where $\wedge(t)=-\int_0^t \ln|2\sin(s)|ds$ is the Lobachevsky function. Rivin proved directly that $V(\theta_1,\theta_2,\theta_3)$ is strictly concave in $E_\theta(2)=\{(\theta_1,\theta_2,\theta_3)\in\mathbb{R}^3_{>0}\,|\,\theta_1+\theta_2+\theta_3=\pi\}$.

The framework of the variational principle is as follows. Define a *corner* in a triangle to be a triple (v,e,f) where v is a vertex, f is a triangle having v as a vertex and e is an edge in f facing v. Let $\mathrm{Cor}(S,T)$ be the set of all corners in (S,T). For a triangulated surface (S,T), there are $3|T^{(2)}|$ corners.

Definition 6.5. *A Euclidean angle structure on a triangulated surface is a map* $\alpha:\mathrm{Cor}(S,T)\to\mathbb{R}_{>0}$ *so that if* c_1,c_2,c_3 *are corners in a triangle, then* $\alpha_1+\alpha_2+\alpha_3=\pi$ *where* $\alpha_i=\alpha(c_i)$. *We call* $\alpha(c)$ *the angle of the corner* c.

Evidently, if (S,T,l) is a Euclidean polyhedral metric on (S,T), then its inner angles define an angle structure on (S,T) called the associated angle structure.

Let the space of all Euclidean angle structures in (S,T) be $A(S,T)$. The ϕ_0-curvature can be naturally defined on $A(S,T)$ by $\phi_0(e)=\pi-a-b$ where a,b are the angles facing the edge e.

Remark 6.6. *The notion of angle structure first appeared in Colin de Verdiere's work [CV1].*

Given any $g:A(S,T)\to\mathbb{R}$, define the subspace $A(S,T,g)=\{x\in A(S,T)\mid$ the ϕ_0 curvature of x is $g\}$. By definition, both $A(S,T)$ and $A(S,T,h)$ are convex polytopes.

Rivin's global rigidity theorem follows from the fact that for any $g:A(S,T)\to\mathbb{R}$, there is at most one Euclidean polyhedral metric on (S,T) whose associated angle structure is in $A(S,T,g)$. To prove the later statement, we define the energy function $W:A(S,T)\to\mathbb{R}$ by

$$W(x)=\sum_{\{e_i,e_j,e_k\}\in T^{(2)}} V(x_i,x_j,x_k)$$

where x_i is the angle of the corner in the triangle $\{e_i,e_j,e_k\}$ facing the edge e_i. The function W is strictly concave in the convex set $A(S,T)$ due to the strictly concavity of the volume function V.

Lemma 6.7. *If* $x\in A(S,T,h)$ *is the angle structure associated to a Euclidean polyhedral metric, then* x *is a critical point of* $W|$.

Since a strictly concave function on a convex set has at most one critical point, Lemma 6.7 implies Theorem 6.4.

Proof (Lemma 6.7). Let $l : E \to \mathbb{R}_{>0}$ be a polyhedral metric whose associated angle structure is x. Let $v \in T_x A(S,T,h)$ be a tangent vector. By definition, $v : \mathrm{Cor}(S,T) \to \mathbb{R}$ so that $v_i + v_j + v_k = 0$ for three angles v_i, v_j, v_k of each triangle and $v_i + v_j = 0$ when v_i and v_j are angles facing an edge. We claim that $\frac{dW}{dt}(x+tv)|_{t=0} = 0$, i.e., x is a critical point of $W|$. Indeed, by definition,

$$\frac{dW}{dt}(x+tv)|_{t=0} = -\sum_{\{e_i,e_j,e_k\}\in T^{(2)}} [\ln(2\sin x_i)v_i + \ln(2\sin x_j)v_j + \ln(2\sin x_k)v_k] \tag{6.1}$$

Since $v_i + v_j + v_k = 0$ and $2\sin x_i = rl_i$ for some constant r depending only on the triangle $\{e_i, e_j, e_k\}$ where l_s is the length of the edge e_s, it follows that (6.1) is equal to

$$= -\sum_{\{e_i,e_j,e_k\}\in T^{(2)}} [\ln(l_i)v_i + \ln(l_j)v_j + \ln(l_k)v_k]$$

$$= -\sum_{e\in E} \ln(l(e))(v_i + v_j)$$

where v_i and v_j are the angles facing the edge e. But by definition of v, $v_i + v_j = 0$. Thus $\frac{dW}{dt}(c+tv)|_{t=0} = 0$, i.e., x is a critical point of $W|$. □

Remark 6.8. *The converse of the lemma is also true. Namely, if x is the critical point of $W| : A(S,T,h) \to \mathbb{R}$, then there exists a Euclidean polyhedral metric whose associated angle structure is x. We leave it as an exercise to the reader. (hint: use the Lagrangian multiplier.) See also [Ri] for a proof.*

7

Polyhedral Surfaces of Circle Packing Type

7.1 Introduction

In his work on constructing hyperbolic metrics on 3-manifolds, W.Thurston introduced a class of polyhedral surfaces of circle packing type. Suppose (S,T) is a closed triangulated surface with V,E as the sets of all vertices and edges. By definition, a Euclidean or hyperbolic circle packing metric on (S,T) is a map, $r : V \to \mathbb{R}_{>0}$, called the *radius parameter* so that the associated Euclidean or hyperbolic polyhedral metric $l : E \to \mathbb{R}_{>0}$ is given by

$$l(uv) = r(u) + r(v)$$

where u, v are the end points of the edge uv. Note that the triangular inequality for l at the three edges of each triangle is automatically satisfied. see Fig. 7.1. Thurston and Andreev proved the following remarkable theorem for circle packing polyhedral surfaces.

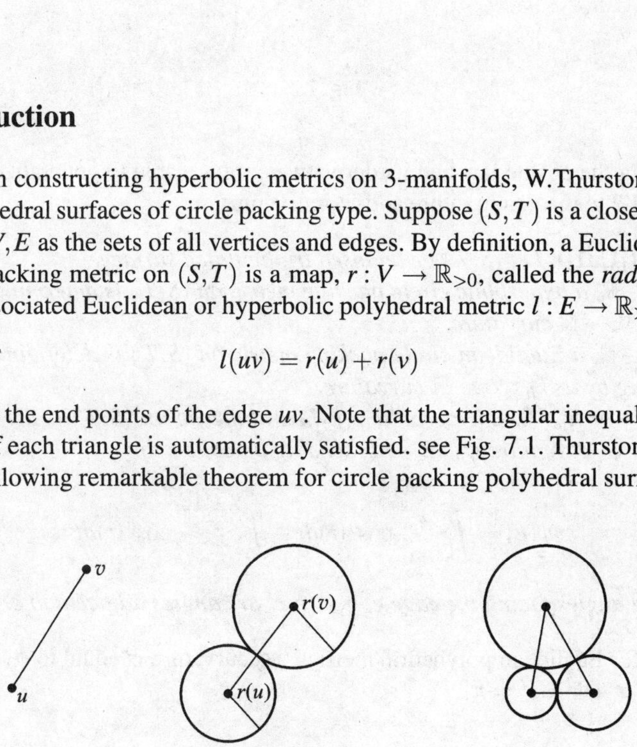

Fig. 7.1.

Theorem 7.1 (Thurston-Andreev). *Let (S,T) be a closed triangulated surface.*
(a) A hyperbolic circle packing metric on (S,T) is determined up to isometry by its discrete curvature $k_0 : V \to \mathbb{R}$. Furthermore, the set of all k_0-curvatures of $\mathbb{H}^2$ circle packing metrics on (S,T) is a convex polytope.
(b) A Euclidean circle packing metric on (S,T) is determined up to isometry and scaling by its discrete curvature $k_0 : V \to \mathbb{R}$. Furthermore, the set of all k_0-curvatures of $\mathbb{E}^2$ circle packing metrics on (S,T) is a convex polytope.

In this section, we will present Colin de Verdiere's proof [CV1] of the rigidity part of Theorem 7.1 based on the variational principle and Marden-Rodin's proof [MaR] of the convexity of the space of discrete curvatures in Theorem 7.1. We will also prove G. Leibon's rigidity theorem for the hyperbolic polyhedral surface in ψ_0 curvature.

Theorem 7.2 (Leibon). *A hyperbolic metric on (S,T) is determined up to isometry by its ψ_0 curvature $\psi_0 : E \to \mathbb{R}$ sending an edge e to $\frac{1}{2}(b+b'+c+c'-a-a')$ where a,a' are angles facing the edge e, b,b',c,c' are angles adjacent to e.*

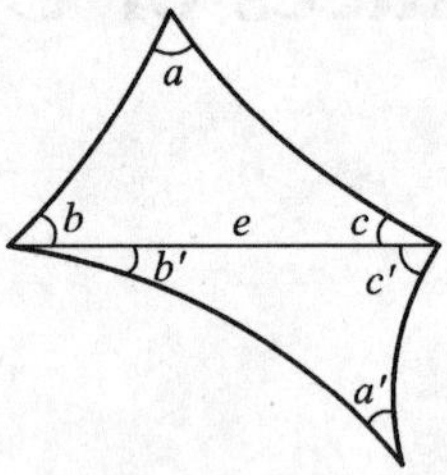

Fig. 7.2.

By studying the cosine law and its derivatives, we are able to generalized both Theorems 7.1 and 7.2 to the k_h-curvature and ψ_h-curvature.

Theorem 7.3 ([Lu1]). *Let (S,T) be a closed triangulated surface.*
(a) *For any $h \in \mathbb{R}$, a hyperbolic circle packing metric on (S,T) is determined up to isometry by its $k_h : V \to \mathbb{R}$ curvature.*
(b) *For any $h \in \mathbb{R}$, a Euclidean circle packing metric on (S,T) is determined up to isometry and scaling by its $k_h : V \to \mathbb{R}$ curvature.*
(c) *For $h \leqslant -1$ or $h \geqslant 0$, a hyperbolic polyhedral metric on (S,T) is determined up to isometry by its ψ_h curvature $\psi_h : E \to \mathbb{R}$ sending an edge e to*

$$\psi_h(e) = \int_0^{\frac{b+c-a}{2}} \cos^h(t)dt + \int_0^{\frac{b'+c'-a'}{2}} \cos^h(t)dt$$

where a,a' are angles facing the edge e, b,b',c,c' are angles adjacent to e.

Note that for Euclidean polyhedral metrics, ψ_0 curvature is equal to ϕ_0 curvature due to $a+b+c=a'+b'+c'=\pi$.

7.2 The Cosine Law and the Radius Parametrization

Given a triangle of edge lengths l_1,l_2,l_3 and opposite angles $\theta_1,\theta_2,\theta_3$, in the Thurston's circle packing approach, one writes $l_i = r_j + r_k, (\{i,j,k\} = \{1,2,3\})$ and consider the relationship between the radii (r_1,r_2,r_3) and angles $(\theta_1,\theta_2,\theta_3)$. In Leibon's approach, one writes $\theta_i = r_j + r_k, (\{i,j,k\} = \{1,2,3\})$ and considers the relationships between (l_1,l_2,l_3) and (r_1,r_2,r_3). The geometric meaning of r_i in Leibon's case is shown below(Fig. 7.3):

It prompts us to consider the cosine law function $(y_1,y_2,y_3) = (y_1(x),y_2(x),y_3(x))$ so that $x_i = r_j + r_k, \{i,j,k\} = \{1,2,3\}$, i.e., $r_i = \frac{x_j+x_k-x_i}{2}$. Thus, we should study the relationship between y and $r = (r_1,r_2,r_3)$. It is summarized in the following lemma. Note that

$$\frac{\partial}{\partial r_i} = \frac{\partial}{\partial x_j} + \frac{\partial}{\partial x_k}.$$

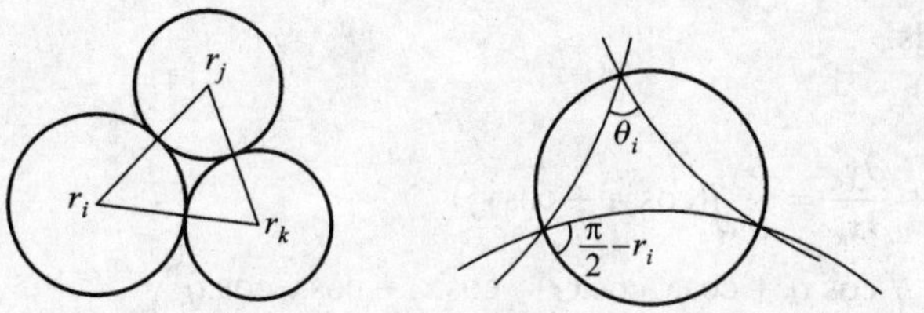

Fig. 7.3.

Lemma 7.4 (The tangent law). *For the cosine law function $y=y(x)$, if $r_i=\frac{1}{2}(x_j+x_k-x_i)$, $\{i,j,k\}=\{1,2,3\}$, then $y=y(r)$ satisfies*

$$\cot^2\frac{y_i}{2}\cos^2 r_i=\cot^2\frac{y_j}{2}\cos^2 r_j \tag{7.1}$$

Furthermore,

$$\frac{1}{\cos r_i}\frac{\partial y_i}{\partial r_j}=\frac{2\cos r_k}{M\sin x_k} \tag{7.2}$$

is symmetric in i, j, and

$$\frac{\partial y_i}{\partial r_i}=\frac{2\sin(x_j+x_k)\cos r_j\cos x_k}{M\sin x_j\sin x_k} \tag{7.3}$$

where M is independent of indices i,j,k.

Remark 7.5. *We call identity (7.1) the tangent law.*

Proof. To see (7.1),

$$\cot^2\frac{y_i}{2}=\frac{1+\cos y_i}{1-\cos y_i}=\frac{\sin x_j\sin x_k+\cos x_i+\cos x_j\cos x_k}{\sin x_j\sin x_k-\cos x_i-\cos x_j\cos x_k}$$
$$=\frac{\cos(x_j-x_k)+\cos x_i}{-\cos(x_j+x_k)-\cos x_i}$$

Using $\cos\alpha_\beta+\cos(\alpha-\beta)=2\cos\alpha\cos\beta$, we see that above is equal to,

$$=(-1)\frac{2\cos r_k\cos r_j}{2\cos r_i\cos(r_1+r_2+r_3)}$$
$$=(-1)\frac{\cos r_i\cos r_j\cos r_k}{\cos^2 r_i\cos(r_1+r_2+r_3)}.$$

The result follows.

Next,

$$\frac{\partial y_i}{\partial r_j}=\frac{\partial y_i}{\partial x_i}+\frac{\partial y_i}{\partial x_k}=\left(\frac{\partial y_i}{\partial x_i}\right)(1+\cos y_j)=\frac{\sin x_i}{A}(1+\cos y_j)$$
$$=\frac{\sin x_i}{A}\frac{1}{\sin x_i\sin x_k}\cdot(\sin x_i\sin x_k+\cos x_j+\cos x_i\cos x_k)$$
$$=\frac{1}{A\sin x_k}\cdot(\cos(x_i-x_k)+\cos x_j)$$
$$=\frac{2}{A\sin x_k}\cos r_k\cos r_i.$$

Thus (7.2) holds.

To see (7.3),

$$\begin{aligned}\frac{\partial y_i}{\partial r_i} &= \frac{\partial y_i}{\partial x_j} + \frac{\partial y_i}{\partial x_k} = \frac{\partial y_i}{\partial x_i}(\cos y_k + \cos y_j) \\ &= \frac{\sin x_i}{A}\left(\frac{\cos x_k + \cos x_i \cos x_j}{\sin x_i \sin x_j} + \frac{\cos x_j + \cos x_i \cos x_k}{\sin x_i \sin x_k}\right) \\ &= \frac{1}{A \sin x_j \sin x_k} \cdot (\sin x_k \cos x_k + \sin x_j \cos x_j + \cos x_i(\cos x_j \sin x_k + \cos x_k \sin x_j)) \\ &= \frac{1}{A \sin x_j \sin x_k}(\sin(x_j + x_k)\cos(x_j - x_k) + \cos x_i \sin(x_j + x_k)) \\ &= \frac{2\sin(x_j + x_k)\cos r_j \cos r_k}{A \sin x_j \sin x_k}.\end{aligned}$$

□

The 2-dimension analogy of the Schlafli formula in the setting of $y = y(r)$ are given by the following theorem. This theorem gives a solution to the basic problem 1 in Chapter 4.

Theorem 7.6. *For the cosine law function $y = y(x)$ with $r = (r_1, r_2, r_3)$ where $r_i = \frac{x_j + x_k - x_i}{2}$, all closed 1-forms of the expression*

$$\sum f(y_i)dg(r_i)$$

for some non-constant smooth functions f, g are, up to scaling and complex conjugation,

$$\eta_h = \sum \left(\int^{y_i} \cot^h \frac{t}{2} dt \right) \cosh^{-1}(r_i) dr_i$$

where $h \in \mathbb{C}$, i.e., $f(t) = \int^t \cot^h \frac{s}{2} ds$ and $g(t) = \int^t \cos^{h-1} s ds$.

Proof. We will verify the closeness, i.e. the following expression is symmetric in i, j

$$\begin{aligned}\frac{\partial}{\partial r_j}\left(\int^{y_i} \cot^h \frac{t}{2} dt \cos^{h-1} r_i\right) &= \cot^h \frac{y_i}{2} \cos^{h-1} r_i \frac{\partial y_i}{\partial r_j} \\ &= \cot^h \frac{y_i}{2} \cos^{h-1} r_i \cdot \cos(r_i) \cdot M \\ &= \left[\cot \frac{y_i}{2} \cos r_i\right]^h \cdot M\end{aligned}$$

where M is symmetric in i, j by the Lemma 7.4. □

Theorem 7.7. *Given a Euclidean or hyperbolic triangle of edge lengths l_i, l_j, l_k and opposite angles $\theta_i, \theta_j, \theta_k$, the following is the complete list, up to scaling, of all closed real-valued 1-forms of the form $\Sigma_{i=1}^3 f(l_i)dg(r_i)$ (where $\theta_i = r_j + r_k$) and $\Sigma_{i=1}^3 f(\theta_i)dg(r_i)$ (where $l_i = r_j + r_k$) for some non-constant smooth functions f, g. Let $h \in \mathbf{R}$ and $u = (u_1, u_2, u_3)$.*

(a) *For a Euclidean triangle of angles θ_i and opposite edge lengths $l_i = r_j + r_k$,*

$$\eta_h = \sum_{i=1}^{3} \frac{\int^{\theta_i} \cot^h(t/2)dt}{r_i^{h+1}} dr_i.$$

Its integration $\int^u \eta_h$ *is locally concave in* $u = (u_1, u_2, u_3)$ *where* $u_i = \int_1^{r_i} t^{-h-1}dt$.

(b) *For a hyperbolic triangle of angles* θ_i *and opposite edge lengths* $l_i = r_j + r_k$,

$$\eta_h = \sum_{i=1}^{3} \frac{\int^{\theta_i} \coth^h(t/2)dt}{\sinh^{h+1} r_i} dr_i.$$

Its integration $\int^u \eta_h$ *is locally strictly concave in* u *where* $u_i = \int_1^{r_i} \sinh^h t dt$.

(c) *For a hyperbolic triangle of edge lengths* l_i *and opposite angles* $\theta_i = r_j + r_k$,

$$\eta_h = \sum_{i=1}^{3} \frac{\int^{l_i} \tanh^h(t/2)dt}{\cos^{h+1} r_i} dr_i.$$

Its integration $\int^u \eta_h$ *is locally strictly convex in* u *where* $u_i = \int_1^{r_i} \cos^{h-1} t dt$.

Proof. The proof of closeness of the 1-forms is just a specialization of Theorem 7.6 by taking care of the curvature factors. For (b), we take $y_i = \pi - \theta_i$ and $x_i = \pi - \sqrt{-1}l_i$ in Theorem 7.6. For (c), we take $y_i = \sqrt{-1}l_i$ and $x_i = \theta_i$. The closeness of the 1-forms together with the convexity in part (a) will be proved below.

To see (a), let $u_i = g(r_i)$ where $g(s) = \int_1^s t^{-h-1}dt$. Then the Hessian of the function $F(u) = \int^u \eta_h$ is the matrix $H = [H_{st}]_{3\times 3} = [\frac{\partial^2 F(u)}{\partial u_t \partial u_s}]$ where $H_{st} = \frac{f'(\theta_s)}{g'(r_t)} \frac{\partial \theta_s}{\partial r_t}$.

We first show that $\eta_h = \Sigma_{i=1}^3 (\int^{\theta_i} \cot^h(t/2)dt) r_i^{-h-1} dr_i$ is closed. By the definition of the radius R of the inscribed circle, we have $\frac{\coth(\theta_i/2)}{r_i} = \frac{1}{R}$. Note that

$$\begin{aligned}
\frac{\partial \theta_i}{\partial r_j} &= \frac{\partial \theta_i}{\partial l_i} + \frac{\partial \theta_i}{\partial l_k} \\
&= \frac{\partial \theta_i}{\partial l_i}(1 - \cos(\theta_j)) \\
&= \frac{2\sin\theta_i \sin^2(\theta_j/2)}{A} \\
&= \frac{4\cot(\theta_i/2)\sin^2(\theta_i/2)\sin^2(\theta_j/2)}{A} \\
&= \frac{4r_i \sin^2(\theta_i/2)\sin^2(\theta_j/2)}{AR}.
\end{aligned} \tag{7.4}$$

By (7.4), for the 1-form $\Sigma_i f(\theta_i)dg(r_i)$ where $f'(\theta) = \cot^h(\theta/2)$ and $g'(r) = r^{-h-1}$, we have

$$\begin{aligned}
\frac{\partial(f(\theta_i)g'(r_i))}{\partial r_j} &= f'(\theta_i)g'(r_i)\frac{\partial \theta_i}{\partial r_j} \\
&= \cot^h(\theta_i/2) r_i^{-h-1} \left(\frac{4r_i}{AR} \sin^2(\theta_i/2)\sin^2(\theta_j/2) \right) \\
&= \frac{4}{AR}(r_i \tan(\theta_i/2))^{-h} \sin^2(\theta_i/2)\sin^2(\theta_j/2) \\
&= \frac{4R^{-h-1}}{A} \sin^2(\theta_i/2)\sin^2(\theta_j/2)
\end{aligned}$$

is symmetric in i, j, i.e., the 1-form $\eta_h = \Sigma_i f(\theta_i)dg(r_i)$ in case (a) is closed.

The Hessian $H = [H_{st}]$ of the integration $F(u) = \int^u \eta_h$ is

$$\begin{aligned} H_{st} &= \frac{f'(\theta_s)}{g'(r_t)}\frac{\partial\theta_s}{\partial r_t} \\ &= (r_s\tan(\theta_s/2))^{-h}(r_sr_t)^h\left(r_t\frac{\partial\theta_s}{\partial r_t}\right) \\ &= R^{-h}(r_sr_t)^h a_{st}. \end{aligned}$$

Since $r_s\tan(\theta_s/2) = R > 0$ is independent of the indices, the above identity shows that the Hessian matrix H is congruent to $[a_{st}]$ where, by (7.4),

$$a_{ij} = r_j\frac{\partial\theta_i}{\partial r_j} = \frac{4r_ir_j}{AR}\sin^2(\theta_i/2)\sin^2(\theta_j/2) > 0 \tag{7.5}$$

and

$$\begin{aligned} a_{ii} &= r_i\frac{\partial\theta_i}{\partial r_i} = r_i\left(\frac{\partial\theta_i}{\partial l_j} + \frac{\partial\theta_i}{\partial l_k}\right) \\ &= -r_i\frac{\partial\theta_i}{\partial l_i}(\cos\theta_k + \cos\theta_j) < 0 \end{aligned} \tag{7.6}$$

due to $\theta_k + \theta_j < \pi$ and $\frac{\partial\theta_i}{\partial l_i} > 0$. Furthermore, we have $|a_{ii}| - |a_{ji}| - |a_{ki}| = -(a_{ii} + a_{ji} + a_{ki}) = -r_i\frac{\partial(\theta_1+\theta_2+\theta_3)}{\partial r_i} = 0$. Thus the matrix $[a_{rs}]$ is semi-negative definite due to the diagonal dominate property above. Thus the Hessian matrix H is semi-negative definite. This shows that the integration is locally concave in u.

Corollary 7.8. *In Theorem 7.7(a), for a Euclidean triangle of edge lengths $l_i = r_j + r_k$, the null space of the Hessian* Hess (F) *where $F(u) = \int^u \eta_h$ at a point u is generated by the vector u for $h \neq 0$ and by $(1,1,1)$ if $h = 0$.*

Proof. Indeed, we first note that the null space of the symmetric 3×3 matrix $[a_{st}]$ contains $(1,1,1)$ due to the equality $a_{ii} + a_{ij} + a_{ik} = 0$. Next, we note that the rank of $[a_{st}]$ is 2, due to the fact that $|a_{ii}| > |a_{ij}|$, for all $i \neq j$. Thus the null space of $[a_{st}]$ is generated by $(1,1,1)$. Now the Hessian matrix H is $R^{-h}[r_s^h r_t^h a_{st}]$ by the calculation above. It follows that the null space of H is generated by $(1,1,1)$ if $h = 0$ or by $\frac{1}{h}(r_1^{-h}, r_2^{-h}, r_3^{-h}) = u$ if $h \neq 0$. □

In Case (b), we take $y_i = \pi - \theta_i$ and $x_i = \pi - \sqrt{-1}l_i$ in Lemma 7.4. By Lemma 7.4 $\cot(\theta_i/2)/\sinh r_i$ is independent of indices and $\frac{1}{\sinh r_i}\frac{\partial\theta_i}{\partial r_j}$ is symmetric in i,j. The Hessian $H = [H_{st}]$ of the integral $F(u) = \int^u \eta_h$ where $f'(\theta) = \cot^h(\theta/2)$ and $g'(r) = \sinh^{-h-1} r$ is

$$\begin{aligned} H_{st} &= \frac{f'(\theta_s)}{g'(r_t)}\frac{\partial\theta_s}{\partial r_t} = \frac{\cot^h(\theta_s/2)}{\sinh^{-h-1}(r_t)}\frac{\partial\theta_s}{\partial r_t} \\ &= \left(\frac{\cot(\theta_s/2)}{\sinh r_s}\right)^h(\sinh r_s\sinh r_t)^h\left(\sinh r_t\frac{\partial\theta_s}{\partial r_t}\right) \\ &= \left(\frac{\cot(\theta_s/2)}{\sinh r_s}\right)^h(\sinh r_s\sinh r_t)^h(\sinh r_t a_{st}). \end{aligned}$$

By the tangent law $\cot(\theta_s/2)/\sinh(r_s) > 0$ is independent of the indices s, the Hessian matrix H of $F(u)$ is congruent to $[a_{st}]$ where

$$
\begin{aligned}
a_{ii} = \sinh r_i \frac{\partial \theta_i}{\partial r_i} &= \sinh r_i \left(\frac{\partial \theta_i}{\partial l_j} + \frac{\partial \theta_i}{\partial l_k} \right) \\
&= -\sinh r_i \frac{\partial \theta_i}{\partial l_i} (\cos\theta_k + \cos\theta_j) \\
&= -\frac{\sinh r_i \sin\theta_i}{A} (\cos\theta_k + \cos\theta_j),
\end{aligned}
$$

and

$$
\begin{aligned}
a_{ij} = \sinh r_j \frac{\partial \theta_i}{\partial r_j} &= \sinh r_j \left(\frac{\partial \theta_i}{\partial l_i} + \frac{\partial \theta_i}{\partial l_k} \right) \\
&= \sinh r_j \frac{\partial \theta_i}{\partial l_i} (1 - \cos\theta_j) \\
&= \frac{2\sinh r_j \sin\theta_i \sin^2(\theta_j/2)}{A}.
\end{aligned}
$$

From the above identities and $\theta_k + \theta_j < \pi$, we see that $a_{ii} < 0$ and $a_{ij} > 0$. Now

$$
\begin{aligned}
&|a_{ii}| - |a_{ji}| - |a_{ki}| \\
&= \frac{\sinh r_i}{A} \left(\sin\theta_i(\cos\theta_k + \cos\theta_j) - 2(\sin\theta_j + \sin\theta_k)\sin^2\frac{\theta_i}{2} \right) \\
&= \frac{4\sinh r_i \sin\frac{\theta_i}{2}}{A} \left(\cos\frac{\theta_i}{2}\cos\frac{\theta_k+\theta_j}{2}\cos\frac{\theta_k-\theta_j}{2} - \sin\frac{\theta_k+\theta_j}{2}\cos\frac{\theta_k-\theta_j}{2}\sin\frac{\theta_i}{2} \right) \\
&= \frac{4\sinh r_i \sin\frac{\theta_i}{2}\cos\frac{\theta_j-\theta_k}{2}}{A} \left(\cos\frac{\theta_i}{2}\cos\frac{\theta_k+\theta_j}{2} - \sin\frac{\theta_i}{2}\sin\frac{\theta_k+\theta_j}{2} \right) \\
&= \frac{4}{A}\sinh r_i \sin\frac{\theta_i}{2}\cos\frac{\theta_j-\theta_k}{2}\cos\frac{\theta_i+\theta_j+\theta_k}{2} > 0.
\end{aligned}
$$

This shows that $-[a_{st}]$ is diagonally dominate. Thus H is negative definite and the integration is locally strictly concave.

In Case (c), we take $y_i = \sqrt{-1}l_i$ and $x_i = \theta_i = r_j + r_k$ in Lemma 7.4. The tangent law says $\tanh(l_i/2)/\cos(r_i)$ is independent of the index i. Furthermore the Hessian $H = [H_{st}]$ of the integral $f(u) = \int^u \eta_h$ where $f'(l) = \tanh^h(l/2)$ and $g'(r) = \cos^{-h-1} r$ is

$$
\begin{aligned}
H_{st} &= \frac{f'(l_s)}{g'(r_t)} \frac{\partial l_s}{\partial r_t} \\
&= \left(\frac{\tanh(l_s/2)}{\cos r_s} \right)^h (\cos r_s \cos r_t)^{h+1} \frac{1}{\cos r_s} \frac{\partial l_s}{\partial r_t} \\
&= \left(\frac{\tanh(l_s/2)}{\cos r_s} \right)^h (\cos r_s \cos r_t)^{h+1} a_{st}.
\end{aligned}
$$

Since $\tanh(l_s/2)/\cos(r_s) > 0$ is independent of the indices, thus the Hessian H is congruent to $[a_{st}]$ where, by Lemma 7.4,

$$a_{ij} = B' \cosh^2(l_i/2)\cosh^2(l_j/2)$$

and

$$a_{ii} = B' \cosh^2(l_i/2)(\cosh^2(l_j/2) + \cosh^2(l_k/2) - 1)$$

for some positive function B' independent of indices. At the equilateral triangle with $\cosh^2(l_i/2) = 2$ for all i, the matrix $[a_{st}]$ is a positive multiple of $[b_{st}]$ where $b_{ij} = 1$ and $b_{ii} = 3/2$. Thus $[a_{st}]$ is positive definite at this triangle. This shows that the matrix H is positive definite for one triangle. But on the other hand, the Hessian H is non-degenerate. Indeed, the Hessian is the Jacobi matrix of the gradient of $F(u) = \int^u \eta_h$ which sends u to $(\int^{u_1} \tanh^h(t/2)dt, \int^{u_2} \tanh^h(t/2)dt, \int^{u_3} \tanh^h(t/2)dt)$. The gradient map is a diffeomorphism since we can solve angle from the length by the cosine law. It follows that $det(H) \neq 0$ for all triangles. Since the space of all hyperbolic triangles is connected and H is symmetric, the signature of H is constant for all hyperbolic triangles. We noticed above that the signature for one triangle is (3,0). It follows that H is positive definite. Thus the integration is locally strictly convex. □

7.3 Colin de Verdiere's Proof of Thurston-Andreev Rigidity Theorem

We will present Colin de Verdiere's proof of the rigidity part of Theorem 7.1(a). Its generalization to any k_h will follow easily due to Theorem 7.6. Furthermore, the proof of Theorem 7.2(b) is similar to that of Theorem 7.2(a) and will be omitted.

Given a hyperbolic triangle of edge lengths $l_i = r_j + r_k$ and apposite angles θ_i where $\{i,j,k\} = \{1,2,3\}$, Colin de Verdiere proved that the 1-form

$$\omega_0 = \sum_{i=1}^{3} \frac{\theta_i}{\sinh(r_i)} dr_i = \sum_{i=1}^{3} \theta_i du_i$$

is closed where $u_i = \int_\infty^{r_i} \frac{1}{\sinh(s)} ds$. (This is the case $h = 0$ in the Theorem 7.6).

Furthermore, the function

$$F_0(u_1,u_2,u_3) = \int_{(-1,-1,-1)}^{(u_1,u_2,u_3)} \omega_0$$

is strictly concave in u so that

$$\frac{\partial F_0}{\partial u_i} = \theta_i, \quad i = 1,2,3.$$

Now, consider the space of all circle packing metrics $r \in \mathbb{R}^V_{>0}$ on (S,T). Let $V = \{v_1, \cdots, v_n\}$. If $f: V \to \mathbb{R}$, we use $f_i = f(v_i)$ and denote $f = (f_1, \cdots, f_n)$. Make a change of variables $u = u(r)$ where

$$u_i = \int_\infty^{r_i} \frac{1}{\sinh t} dt.$$

Then the diffeomorphism $u = u(r)$ sends $\mathbb{R}^V_{>0}$ onto $\mathbb{R}^V_{<0}$, and u will be consider as a new parametrization of the space of all circle packing metrics on (S,T).

Define an energy function $W : \mathbb{R}^V_{<0} \to \mathbb{R}$ by

$$W(u) = \sum_{\{v_i,v_j,v_k\} \in T^{(2)}} F_0(u_i,u_j,u_k)$$

where the sum is over all triangles whose vertices are v_i, v_j, v_k. By the construction, W is strictly concave in $\mathbb{R}^V_{<0}$ since each F_0 is it.

Furthermore,

$$\frac{\partial W}{\partial r_i} = \theta_1 + \cdots + \theta_m = 2\pi - k_0(v_i)$$

where $\theta_1, \cdots, \theta_m$ are angles at the i-th vertex v_i by the basic properties of F_0. By Lemma 5.4, $\nabla W : \mathbb{R}^V_{<0} \to \mathbb{R}^V$ is a smooth embedding. Thus Thurston-Andreev's rigidity theorem follows.

The above proof works for k_h curvature due to Theorem 7.6. We use the energy function F_h instead of F_0. The only attention we should pay to is the convexity of the domain of the function W associated to h. In this case, it is an open cube of the form I^V where $I \subset \mathbb{R}_{<0}$ is an open interval. (To be precisely $I = (-\int_0^\infty \frac{1}{\sin^{h+1} t} dt, 0)$).

7.4 A Proof of Leibon's Theorem

We follow Leibon's original proof of Theorem 7.2. Recall that a corner in a triangulated surface (S,T) is a triple (v,f,e) where v is a vertex, f is a triangle having v as a vertex, and e is the edge in f facing v.

Definition 7.9. *Let (S,T) be a triangulated closed surface so that $\mathrm{Cor}(S,T)$ is the set of all corners in (S,T). A hyperbolic angle structure α on (S,T) is a map $\alpha : \mathrm{Cor}(S,T) \to \mathbb{R}_{>0}$ so that $\alpha(c_i)+\alpha(c_j)+\alpha(c_k) < \pi$ where c_i, c_j, c_k form corners of a triangle. The ψ_0 curvature of a hyperbolic angle structure is defined to be $\psi_0 : E \to \mathbb{R}$ sending an edge e to $b+b'+c+c'-a-a'$ where a,a' are the angles facing e, b,b',c,c' are angles adjacent to e.*

Evidently a hyperbolic polyhedral metric on (S,T) produces a hyperbolic angle structure, called the associated angle structure, by measuring inner angles.

Let $A_1(S,T)$ be the space of all hyperbolic angle structures on (S,T). If $\psi : E \to \mathbb{R}$ is given, we use $A_1(S,T,\psi)$ to denote those hyperbolic angle structures on (S,T) whose ψ_0-curvature is ψ. Evidently both $A_1(S,T)$ and $A_1(S,T,\psi)$ are convex polytopes.

By Theorem 7.6(c), given a hyperbolic triangle of angles $\theta_i = r_j + r_k, \{i,j,k\} = \{1,2,3\}$ and edge lengths l_i, Leibon showed that (for $h=-1$),

$$\eta_1 = \sum_{i=1}^{3} \left(\int^{l_i} \tanh^{-1} \frac{t}{2} dt \right) dr_i = \sum_{i=1}^{3} 2 \ln \sinh \frac{l_i}{2} dr_i$$

is closed and its integration $F_1(r_1,r_2,r_3) = \int_{(\frac{1}{2},\frac{1}{2},\frac{1}{2})}^{(r_1,r_2,r_3)} \eta_{-1}$ is strictly convex in (r_1,r_2,r_3).

For a hyperbolic angle structure $\alpha \in A_1(S,T)$, define its energy function

$$W(\alpha) = \sum_{\sigma=\{i,j,k\} \in T^{(2)}} F_1\left(\frac{\alpha_i+\alpha_j-\alpha_k}{2}, \frac{\alpha_k+\alpha_j-\alpha_i}{2}, \frac{\alpha_i+\alpha_k-\alpha_j}{2} \right)$$

where the sum is over all triangles with three angles $\alpha_i, \alpha_j, \alpha_k$.

Let us make a linear change of variables for $A_1(S,T)$. If $\alpha \in A_1(S,T)$, its r coordinate, $r : \mathrm{Cor}(S,T) \to \mathbb{R}$ is defined to be $r(c) = \frac{1}{2}(\alpha(c')+\alpha(c'')-\alpha(c))$ where c,c',c'' are three corners in the triangle containing c. In particular, $\alpha(c) = r(c')+r(c'')$. Thus, the linear map $\alpha \to r$ is invertible. Let the images $A_1(S,T)$ and $A_1(S,T,\psi)$ under this map be the convex polytopes $B(S,T)$ and $B(S,T,\psi)$. Note that by definition $\psi_0(e) = r(c)+r(c')$ where c,c' are the two corners facing the edge e. We still use $W : B(S,T) \to \mathbb{R}$ to denote the composition $W(r)$ which is

$$W(r) = \sum_{\{v_i,v_j,v_k\}\in T^{(2)}} F_1(r_i,r_j,r_k)$$

where the sum is over all triangles σ containing i-th, j-th, k-th corners, r_i is the value of the function r at the i-th corner. By definition, W is strictly convex in the convex set $B(S,T)$. Furthermore, suppose the i-th corner faces an edge e in a triangle σ containing i-th, j-th, and k-th corner. Then,

$$\frac{\partial W}{\partial r_i} = 2\ln\sinh\frac{l(e)}{2}$$

where $l(e)$ is the length of the edge facing the angle $r_j + r_k$ in a hyperbolic triangle with inner angles $r_j + r_i, r_k + r_i$, and $r_j + r_k$.

Lemma 7.10. *The function $W : B(S,T) \to \mathbb{R}$ is strictly convex and the critical points of $W| : B(S,T,\psi) \to \mathbb{R}$ are the hyperbolic angle structures associated to hyperbolic polyhedral metrics.*

Assuming the lemma, the rigidity theorem of Leibon follows since a strictly convex function defined on a convex set cannot have more than one critical point.

Proof. The convexity of W is evident from the definition. If $r \in B(S,T,\psi)$ is a critical point of $W|$ on $B(S,T,\psi)$, consider an edge $e \in E$ and let c_i and c_j be the two corners facing e, where $i < j$. Then for small t,

$$r(t) = (r_1,\cdots,r_{i-1},r_i+t,r_{i+1},\cdots,r_{j-1},r_j-t,r_{j+1},\cdots,r_n)$$

is a smooth path in $B(S,T,\psi)$ through $r = r(0)$. By the definition of critical points,

$$\frac{d}{dt}|_{t=0}W(r(t)) = 0.$$

By the definition of W and F_1, we obtain

$$\ln\sinh\frac{l(e)}{2} - \ln\sinh\frac{l'(e)}{2} = 0$$

where $l(e)$ and $l'(e)$ are the two lengths of the edge e in the two hyperbolic triangles adjacent to e whose inner angles are given by $\alpha(r)$. This shows $l(e) = l(e')$, i.e., there exists a hyperbolic polyhedral metric whose associated angle structure is $\alpha = \alpha(r)$. Conversely, the result follows from the Lagrangian multipliers. In our case, a direct argument is also possible. For $r \in B(S,T,\psi)$, its tangent space $T_rB(S,T,\psi)$ is generated by vectors of the form $\frac{d}{dt}r(t)$ for edges e running over all elements in E. Thus, if r is associated to a hyperbolic polyhedral metrics, then $\frac{d}{dt}|_{t=0}W(r(t)) = 0$ for all $e \in E$. This implies that $\frac{d}{dt}|_{t=0}W(\rho(t)) = 0$ for all smooth paths ρ through r in $B(S,T,\psi)$. This shows that r is a critical point. □

7.5 A Sketch of a Proof of Theorem 7.3(c)

The proof of Theorem 7.3(c) for $h \geqslant 0$ follows the same principle used in §7.4 by using the convex energy function

$$F_h(u_1,u_2,u_3) = \int^{(l_1,l_2,l_3)} \eta_h$$

where $\eta_h = \sum_{i=1}^{3} \left(\int_1^{l_i} \cos^{h+1} \frac{t}{2} dt \right) du_i$ with $u_i = \int_0^{r_i} \cosh^h t dt$. By definition, we have

$$\frac{\partial F_h}{\partial u_i} = \int_1^{l_i} \cos^{h+1} \frac{t}{2} dt.$$

For a hyperbolic angle structure α on (S,T), define its h-th radius coordinate $u: V \to \mathbb{R}$ to be

$$u(c) = \int_0^{\frac{1}{2}(\alpha(c')+\alpha(c'')-\alpha(c))} \cosh^h t dt$$

where c', c'', c are corners of a triangle. Since the map $f(t) = \int_0^t \cos^h(s) ds$ is a diffeomorphism from $(-\frac{\pi}{2}, \frac{\pi}{2})$ onto an interval, the map $\alpha \mapsto r_\alpha$ is a diffeomorphism sending the convex set $A_1(S,T)$ onto a domain $X(S,T)$. We will prove below that for $h \geqslant 0$, $X(S,T)$ is an open convex set.

Assuming this, then Theorem 7.3(c) follows by essentially the same argument as in §7.4. Define a map

$$W: X(S,T) \to \mathbb{R}$$

by

$$W(u_1, \cdots, u_n) = \sum_{\sigma=\{i,j,k\} \in T^{(2)}} F_h(u_i,u_j,u_k)$$

where the sum is over all triangles σ whose corners are c_i, c_j, c_k and

$$u_i = \int_0^{\frac{1}{2}(\alpha(c_j)+\alpha(c_k)-\alpha(c_i))} \cosh^h s ds.$$

By Theorem 7.6, W is a strictly convex function defined on an open convex set $X(S,T)$. For any $\psi: E \to \mathbb{R}$, define $X(S,T,\psi) = \{u \in X(S,T) | u(c) + u(c') = \psi(e)\}$ where c, c' are two corners facing an edge e.

By definition, $X(S,T,\psi)$ is convex. Now, the same argument as in Lemma 7.10 shows that the critical points of $W|$ on $X(S,T,\psi)$ are exactly those induced by a hyperbolic polyhedral metrics on (S,T). Thus, theorem 7.3(b) follows.

It remains to prove the convexity of $X(S,T)$ when $h \geqslant 0$.

By definition $X(S,T)$ is the product $\Omega \times \Omega \times \cdots \times \Omega$ where Ω is the image of $\{r = (r_1,r_2,r_3) \in \mathbb{R}^3 | r_i + r_j > 0, r_1 + r_2 + r_3 < \frac{\pi}{2}, i \neq j\}$ under the diffeomorphism $u = u(r)$ where $u = (u_1,u_2,u_3)$ and $u_i = p(r_i)$ with $p(t) = \int_0^t \cosh^h(s) ds$ in the interval $(-\frac{\pi}{2}, \frac{\pi}{2})$. Thus, it suffices to show that Ω is convex in $\mathbb{R}^3$.

To this end, let us first note that $p(-t) = -p(t)$ and $p'(t) > 0$, thus $r_i + r_j > 0$ if and only if $u_i + u_j > 0$. This shows that

$$\Omega = \left\{ (u_1,u_2,u_3) \in T^3 \middle| u_i + u_j > 0, u_3 < p\left(\frac{\pi}{2} - g(u_1) - g(u_2)\right) \right\}.$$

Now the convexity of Ω for $h \geqslant 0$ follows from the same calculation as in Chapter 6 using Lemma 6.3.

7.6 Marden-Rodin's Proof Thurston-Andreev Theorem

For a triangulated closed surface (S,T), Thurston-Andreev's Theorem 7.1 asserts that the space of all discrete curvatures $\{k_0\}$ of Euclidean (or hyperbolic) circle packing metrics on (S,T) is a convex polytope. We present Marden-Rodin's elegant proof [MaR] of it in this section. For simplicity, we deal with the hyperbolic metrics only. The same argument also works for Euclidean case without too much changes.

To begin with, let V,E be the sets of all vertices and edges in the triangulation T. A hyperbolic circle packing metric on (S,T) is denote by its radius parameter, $r \in \mathbb{R}^V_{>0}$. The curvature of r is denoted by $k_0 : V \to \mathbb{R}$. Let $\Phi : \mathbb{R}^V_{>0} \to \mathbb{R}^V$ be the curvature map sending r to k_0. The goal is to prove

Proposition 7.11 (Thurston). *For hyperbolic circle packing metrics on (S,T), the set of all discrete curvature $K(\mathbb{R}^V_{>0})$ is the open convex polytope*

$$\Omega = \{z \in \mathbb{R}^V \,|\, inequvalities\ (7.7), (7.8)\ hold\ \}$$

where

$$z(v) < 2\pi, \forall v \in V, \tag{7.7}$$

and for any subset $I \subset E$, let F_I be the set of all triangles in T having a vertex in I,

$$\sum_{v\in I} z(v) > 2\pi|I| - \pi|F_I|. \tag{7.8}$$

Proof. (Marden-Rodin) By the rigidity theorem of Thurston-Andreev, the map $\Phi : \mathbb{R}^V_{>0} \to \mathbb{R}^V$ is a smooth embedding. Thus $\Phi(\mathbb{R}^V_{>0})$ is open in $\mathbb{R}^V$. We will show that $\Phi(\mathbb{R}^V_{>0})$ is an open subset of Ω and $\Phi(\mathbb{R}^V_{>0})$ is closed in Ω. Since Ω is connected due to convexity, it follows that $\Phi(\mathbb{R}^V_{>0}) = \Omega$.

We begin with a study of degeneration of hyperbolic triangles. Suppose a hyperbolic triangles has edge lengths r_1+r_2, r_2+r_3 and r_3+r_1, so that the angle opposite to (r_i+r_j)-th edge is θ_k where $\{i,j,k\} = \{1,2,3\}$. We say a sequence of hyperbolic metrics degenerates if one of r_i converges to 0 or ∞.

Lemma 7.12 ([MaR] and [Thu]). *Under the assumption above*
(a) $\lim_{r_i\to\infty} \theta_i(r_1,r_2,r_3) = 0$ *so that the convergence is uniform,*
(b) *Suppose $(f_i,f_j,f_k) \in \mathbb{R}^3_{>0}$. Then for $\{i,j,k\} = \{1,2,3\}$,*

$$\lim_{(r_i,r_j,r_k)\to(0,f_j,f_k)} \theta_i(r_1,r_2,r_3) = \pi,$$

$$\lim_{(r_i,r_j,r_k)\to(0,0,f_k)} \theta_i(r_1,r_2,r_3) = \pi,$$

$$\lim_{(r_i,r_j,r_k)\to(0,0,0)} \theta_1+\theta_2+\theta_3 = \pi.$$

Proof. To see (a), after applying an isometry, we may assume that the i-th vertex is at the center of the Poincare disk model of the hyperbolic space. For large radius r_i, the hyperbolic ball of radius r_i centered at the origin is a Euclidean ball centered at the origin whose Euclidean radius is almost one. Thus, no matter how r_j and r_k are chosen, the j-th and k-th vertices are close to the unit circle $|z| = 1$ and the Euclidean distance between them is very small. This implies the angle θ_i is arbitrary small if r_i is large since the Euclidean and hyperbolic measurements of angles at the i-th vertex are the same. See Fig. 7.4.

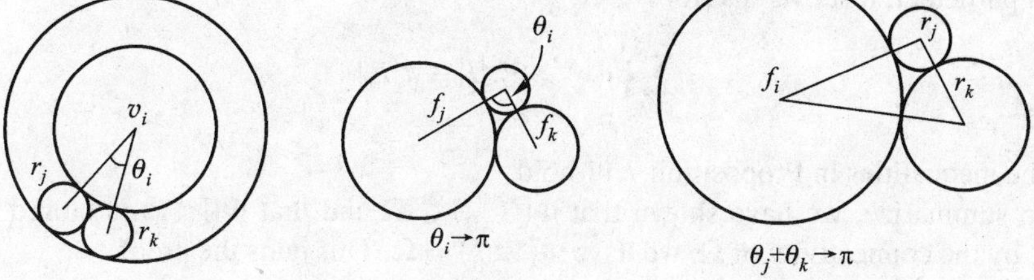

Fig. 7.4.

To see part (b), note the last limit follows from the Gauss-Bonnet theorem. Indeed, since (r_1, r_2, r_3) converges to 0, the diameter of the triangle tends to zero. Thus, its area converges to 0.

The other two inequalities follows from the same geometric argument. See Fig. 7.4.□

Using this lemma, we will identify the boundary of $\Phi(\mathbb{R}^V_{>0})$ as follows. Take a sequence of metrics $r^{(m)} \in \mathbb{R}^V_{>0}$ so that $\lim_{m\to\infty} r^{(m)} = a \in [0,\infty]^V$ where $a(v) = 0$ or ∞ for some $v \in V$. Let us consider the sequence of vectors $\Phi(r^{(m)})$,

Case 1. If there is $v \in V$ so that $a(v) = +\infty$, then using Lemma 7.12(a) above to all triangles having v as a vertex, we conclude that the sum of all inner angles at v converges to 0. Thus

$$\lim_{m\to\infty} k_0(v) = 2\pi$$

However, from the definition, $k_0(v) < 2\pi$, i.e., (7.7) holds. This shows $\Phi(\mathbb{R}^V_{>0}) \subset \{z \in \mathbb{R}^V | z(v) < 2\pi\}$ and $\lim \Phi(l^{(m)})$ is in the hyperplane $z(v) = 2\pi$.

Case 2. If $a \in [0,\infty)^V$ and there is $v \in V$ so that $a(v) = 0$, let I be the non-empty set $\{v \in V | a(v) = 0\}$, and F_I be the set of all triangles having a vertex in I. Take a triangle σ in F_I. Let the vertices of σ be v_1, v_2, v_3 and angles at these vertices $\theta_1, \theta_2, \theta_3$. There are three possibilities for σ: (i) there is only one vertex, say $v_1 \in I$; (ii) there are exactly two vertices, say v_1, v_2 in I; and (iii) all vertices v_1, v_2, v_3 are in I. By Lemma 7.12(b), in the case (i) $\lim_{m\to\infty} \theta_1 = \pi$. In the case (ii), $\lim_{m\to\infty}(\theta_1 + \theta_2) = \pi$ and in the case (iii) $\lim_{m\to\infty}(\theta_1 + \theta_2 + \theta_3) = \pi$.

Now consider the sum of all discrete curvatures at vertices in I,

$$\sum_{v\in I} k_0(v) = 2\pi|I| - \sum_{\theta\in X} \theta$$

where X is the set of all inner angles having vertices in I. We will group inner angles in X according to the triangles in F_I in which they lie. If we sum over all inner angles in X which lie in a single $\sigma \in F_I$, by Lemma 7.12(b), the sum converges to π.

It follows that

$$\lim_{m\to\infty} \sum_{\theta\in X} \theta = \pi|F_I|.$$

This shows that

$$\lim_{m\to\infty} \sum_{v\in I} k_0(v) = 2\pi|I| - \pi|F_I|.$$

Furthermore, since the sum of inner angles of a hyperbolic triangle is less than π, it follows that $\sum_{\theta\in X} \theta < \pi|F_I|$ for any $r \in \mathbb{R}^V_{>0}$.

In particular, it shows that for $r \in \mathbb{R}^V_{>0}$

$$\sum_{v \in I} k_0(v) > 2\pi|I| - \pi|F_I|$$

i.e., the inequalities in Proposition 7.11 hold.

To summarize, we have shown that $\Phi(\mathbb{R}^V_{>0}) \subset \Omega$ and that $\Phi(\mathbb{R}^V_{>0})$ is closed in Ω. Thus, by the connectivity of Ω, we have $\Phi(\mathbb{R}^V_{>0}) = \Omega$. This ends the proof. □

8

Non-negative Curvature metrics and Delaunay Polytopes

We will establish some relationships between those non-negative ϕ_h and ψ_h curvature polyhedral metrics with the Delaunay condition. The following convention will be used throughout the chapter. Let K^2 be $\mathbb{E}^2$ or $\mathbb{S}^2$ or $\mathbb{H}^2$. Let (S,T) be a closed triangled surface with $V,E,T^{(2)}$ the sets of all vertices, edges and triangles. The space of all K^2 polyhedral metrics on (S,T) is denoted by $P_{K^2}(S,T)$ and is considered as a subset of $\mathbb{R}^E_{>0}$ by the length parametrization.

Most of the results in this chapter are due to I.Rivin and G.Leibon.

8.1 Non-negative ϕ_h and ψ_h Curvature Metrics and Delaunay Condition

For $h \in \mathbb{R}$ recall that ψ_h and ψ_h curvatures of a K^2 polyhedral metric are maps from E to $\mathbb{R}$ given by

$$\phi_h(e) = \int_a^{\frac{\pi}{2}} \sin^h t dt + \int_{a'}^{\frac{\pi}{2}} \sin^h t dt$$

and

$$\psi_h(e) = \int_0^{\frac{b+c-a}{2}} \cos^h t dt + \int_0^{\frac{b'+c'-a'}{2}} \cos^h t dt$$

where a,a' are angles facing the edge e and b,b',c,c' are angles adjacent to e. The following result says positivity of ϕ_h and ψ_h curvature is independent of the choice of $h \in \mathbb{R}$. Note that if $K^2 = \mathbb{E}^2$, then $\phi_0(e) = \psi_0(e)$ due to $a+b+c = a'+b'+c' = \pi$.

Proposition 8.1. (1) $\psi_h(e) \geqslant 0$ *if and only if* $\psi_0(e) \geqslant 0$.
(2) $\phi_h(e) \geqslant 0$ *if and only if* $\phi_0(e) \geqslant 0$.

Proof. Consider the function $g(x) = \int_0^x \cos^h t dt, t \in (-\frac{\pi}{2}, \frac{\pi}{2})$. It is an odd function $g(-x) = g(x)$ so that $g'(x) > 0$. It follows that $g(x) + g(y) \geqslant 0$ if and only if $g(x) \geqslant g(-y)$. This is the same as $x \geqslant -y$, i.e., $x+y \geqslant 0$. Now $\psi_h(e) = g(\frac{b+c-a}{2}) + g(\frac{b'+c'-a'}{2})$. Thus $\psi_h(e) \geqslant 0$ if and only if $\frac{b+c-a}{2} + \frac{b'+c'-a'}{2} \geqslant 0$, i.e., $\psi_0(e) \geqslant 0$. This establishes (1). To see (2), note that $\int_a^{\frac{\pi}{2}} \sin^h t dt = \int_0^{\frac{\pi}{2}-a} \cos^h t dt = g(\frac{\pi}{2} - a)$. By the same argument, part (2) follows. □

Now we will relate positivity of ψ_0 or ϕ_0 with the Delaunay condition. Let us begin with the simplest case of a Delaunay decomposition. Given a finite set of at least two points V in the Euclidean plane $\mathbb{E}^2$, for each $v \in V$, consider the Voronoi domain $D(v)$

associated to v, $D(v)=\{x\in\mathbb{R}^2|d(x,v)\leqslant d(x,u),\forall u\in V\}$. By definition $D(v)$ is a convex polytope which is the intersection of a finite set of half-planes. Furthermore, if $v\neq v'$, then the intersection $D(v)\cap D(v')$ is either empty or a line segment or a point. The triple intersection $D(v)\cap D(v')\cap D(v'')$ for three distinct vertices is either a point or is empty. This shows that the Voronoi domains produce a geometric cell decomposition $\widetilde{T}$ of the plane $\mathbb{E}^2$. The vertices of $\widetilde{T}$ are points of the shortest equal distance to at least three points in V. The Delaunay decomposition of $\mathbb{E}^2$ with vertices V is the dual cell decomposition to $\widetilde{T}$. To be precise, vertices of T are points in V. If $D(v)$ and $D(v')$ are two Voronoi domains sharing an edge in $\widetilde{T}$, draw the edge vv' in T. The 2-cells in T correspond bijectively with vertices of $\widetilde{T}$ so that if $u\in\widetilde{T}^{(0)}$ is a vertex of the shortest equal distance to $v_{i_1},\cdots,v_{i_n}$, then the dual 2-cell in $T^{(2)}$ has vertices $v_{i_1},\cdots,v_{i_n}$. By the construction, the vertices $v_{i_1},\cdots,v_{i_n}$ lie in the circle centered at u so that the interior of the circle contains no other points in V. For generic choice of points in V, the compact 2-cells in the Delaunay decomposition T are triangles. In this case, we call it a Delaunay triangulation (or geometric triangulation satisfying the Delaunay condition). The equivalent way to say is that the circumference circle of each triangle contains no vertex in its interior. Let us call it the circumcircle condition for geometric triangles in K^2 geometry.

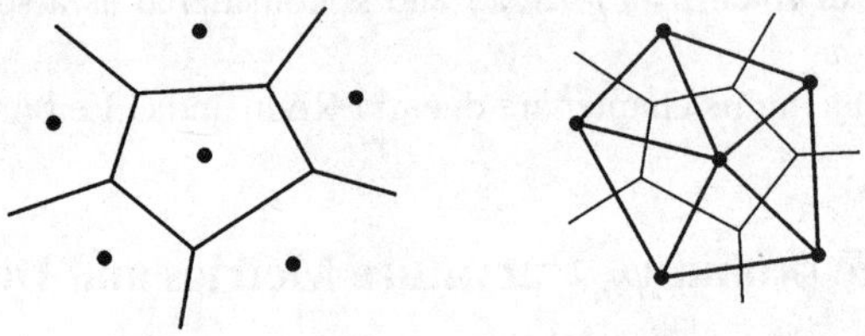

Fig. 8.1.

Lemma 8.2. *A geometric triangulation T of a region in $\mathbb{E}^2$ or $\mathbb{H}^2$ or $\mathbb{S}^2$ satisfies the circumcircle condition if and only if for each edge e in T adjacent to two triangles, $\psi_0(e)\geqslant 0$.*

Proof. The simplest case is two Euclidean triangles $\triangle ABC$ and $\triangle ABD$ in $\mathbb{E}^2$ sharing an edge AB. The vertex D lies outside the circumcircle of $\triangle ABC$ if and only if $\angle C+\angle D\leqslant\pi$, i.e., $\phi_0(e)\geqslant 0$. In the general case, a spherical or a hyperbolic triangle is a triangle $\triangle$ in $\mathbb{E}^2$ bounded by three circular arcs of inner angles $\theta_1,\theta_2,\theta_3$.

Let X be the circle passing through all three vertices of the triangle $\triangle$. Then the angles of the intersection of the circumcircle X with three edges of $\triangle$ are $\frac{\pi}{2}-r_i$ where $r_i=\frac{1}{2}(\theta_j+\theta_k-\theta_i),\{i,j,k\}=\{1,2,3\}$. Thus the non-vertex circumference condition is

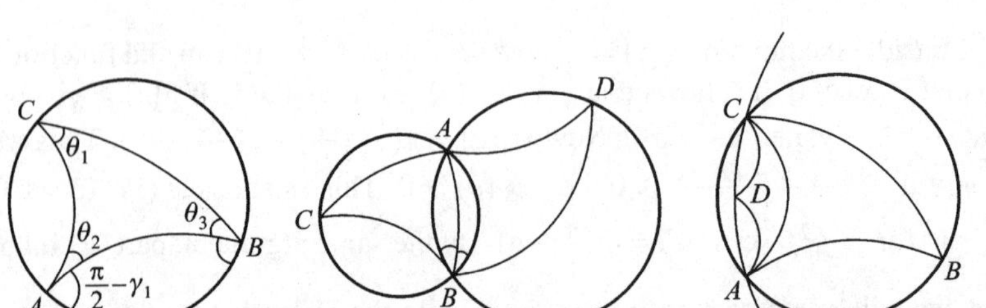

Fig. 8.2.

equivalent to say that $(\frac{\pi}{2}-\frac{1}{2}(b+c-a))+(\frac{\pi}{2}-\frac{1}{2}(b'+c'-a'))\leqslant\pi$, i.e., $\psi_0(e)\geqslant 0$. (see Fig. 8.2) □

This prompts the following

Definition 8.3. *A K^2 polyhrdral surface (S,T,d) is called Delaunay if $\psi_0(e)\geqslant 0$ for all edges e.*

Note in the case of Euclidean polyhedral surface, $\psi_0(e)=\phi_0(e)\geqslant 0$. The condition $\phi_0(e)\geqslant 0$ first appeared in Rivin's work [Ri] in which he established a relationship between $\phi_0(e)\geqslant 0$ and convex polyhedra with vertices at the sphere at infinity of the hyperbolic 3-space. Here is Rivin's beautiful observation. Suppose P is such a convex polyhedron so that all it faces are triangles. Take one vertex of P to be the ∞ in the upper-half-space model $\mathbb{H}^3$ of the hyperbolci 3-space. Let Q be the orthogonal projection of P in the plane $\mathbb{E}^2$ (see Fig.8.3). Consider all ideal tetrahedra in P formed by the vertex ∞ and triangle faces in P. These ideal tetrahedra correspond to Euclidean triangles in $\mathbb{E}^2$ by projection. These triangles produce a geometric triangulation of Q. If e is an edge in the triangulation of Q adjacent to two triangles $\triangle_1,\triangle_2$, then $\phi_0(e)=\pi-a-b$ is the exterior dihedral angle of P at the edge which projects to e. In summary, Delaunay-conditions and the convexity of Q are equivalent to the convexity of the 3-dimensional ideal hyperbolic polyhedral.

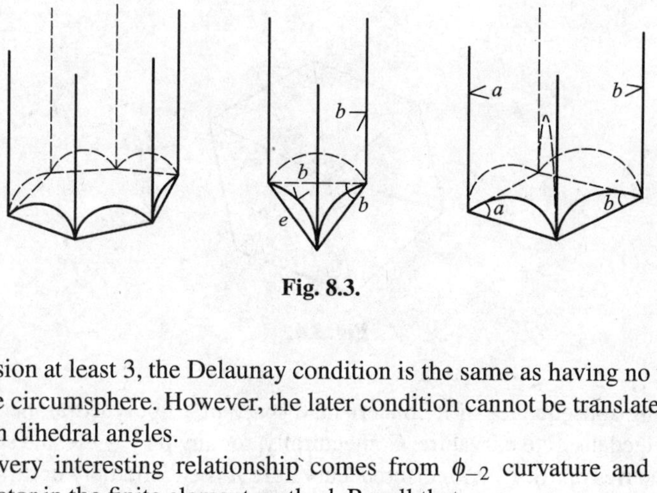

Fig. 8.3.

In dimension at least 3, the Delaunay condition is the same as having no vertex in the interior of the circumsphere. However, the later condition cannot be translated into linear inequalities in dihedral angles.

Another very interesting relationship comes from ϕ_{-2} curvature and the discrete Laplace operator in the finite element method. Recall that

$$\phi_{-2}(e)=\cot a+\cot a'$$

where a,a' are angles facing the edge e. If (S,T,d) is an Euclidean polyhedral surface, then the discrete Laplacian operator $\triangle:\mathbb{R}^V\to\mathbb{R}^V$ is given by the formula

$$\triangle(f)(v)=\sum_{uv\in E}\phi_{-1}(uv)(f(u)-f(v)).$$

It is an interesting question to find if there is discrete analogy of Weyl's formula for eigenvalues of the Laplace operators, relating eigenvalues with geometry of the surface.

8.2 Relationship between ϕ_0, ψ_0 Curvature and the Discrete Curvature k_0

The work of Rivin and Leibon on ϕ_0 and ψ_0 curvatures shows that one can recover k_0 curvature from ϕ_0 and ψ_0.

Lemma 8.4 ([Le], [Ri]). *Suppose v is a vertex in the triangulated closed surface (S,T) so that $e_1,\cdots,e_k$ are all edges adjacent to v (counted with multiplicity). Then for all K^2 polyhedral metrics on (S,T),*

$$\sum_{i=1}^{k}\psi_0(e_i)=2\pi-k_0(v).$$

In particular, for a Euclidean polyhedral surface, $\sum_{i=1}^{k}\phi_0(e_i)=2\pi-k_0(v)$.

Indeed, let us label the set of all triangles adjacent to v cyclically by $\sigma_1,\cdots,\sigma_k$ so that σ_i has two edges e_i and e_{i+1}. Let the angles in σ_i be a_i,b_i,c_i so that b_i and c_i face e_{i+1} and e_i respectively. (see Fig 8.4). Then

$$\psi_0(e)=\sum_{i=1}^{k}\left(\frac{1}{2}(a_i+b_i-c_i)+\frac{1}{2}(a_{i+1}+c_{i+1}-b_{i+1})\right)=\sum_{i=1}^{k}a_i=2\pi-k_0(v).$$

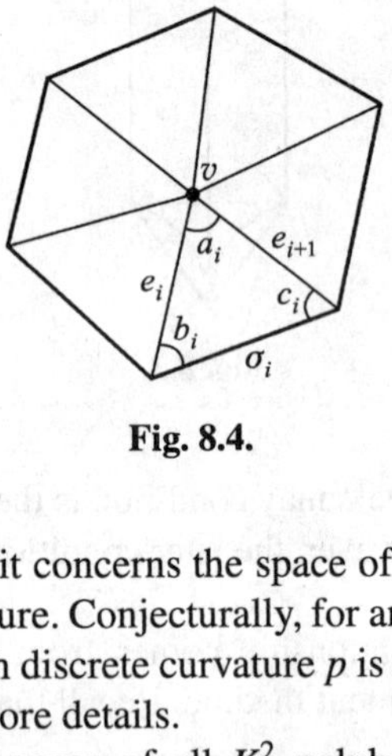

Fig. 8.4.

One interesting consequence of it concerns the space of all polyhedral metrics on (S,T) with a prescribed discrete curvature. Conjecturally, for any $p:V\to\mathbb{R}$, the space of all K^2 polyhedral metrics on (S,T) with discrete curvature p is either empty or diffeomorphic to $\mathbb{R}^N$ for some N. See [Lu1] for more details.

Recall that $P_{K^2}(S,T)$ is the space of all K^2 polyhedral metrics on (S,T). Let π : $P_{K^2}(S,T)\to\mathbb{R}^V$ be the discrete curvature map sending a metric to its discrete curvature k_0.

Proposition 8.5. (a) *For a hyperbolic polyhedral surface, $\pi:P_{\mathbb{H}^2}(S,T)\to\mathbb{R}^V$ is a submersion.*

(b) *For a Euclidean polyhedral surfaces, $\pi:P_{\mathbb{E}^2}(S,T)\to\{z\in\mathbb{R}^V\,|\,\sum_{v\in V}z(v)=2\pi\chi(S)\}$ is a submersion.*

Note that for Euclidean polyhedral surfaces $\sum_{v\in V}z(v)=2\pi\chi(S)$ is the Gauss-Bonnet formula. In particular, in the two cases above, the space $\pi^{-1}(p)$ of polyhedral metrics with prescribed discrete curvature is a smooth manifolds.

Proof. We will give a proof of part (a). The same argument also works for (b). The main idea is to use Leibon's theorem that $\Psi_0 : P_{\mathbb{H}^2}(S,T) \to \mathbb{R}^E$ sending a metric to its ψ_0 curvature is a smooth embedding. Let $G : \mathbb{R}^E \to \mathbb{R}^V$ be the linear map defined by Lemma 8.4, i.e., if $z \in \mathbb{R}^E$ then $G(z)(v) = \sum_{i=1}^k z(e_i)$ for each $v \in V$ and all edges $e_1, \cdots, e_k$ adjacent to v. By definition and lemma 8.4, $\pi = -G \circ \Psi_0 + 2\pi(1, \cdots, 1)$. By Leibon's rigidity theorem that Ψ_0 is a submersion, it suffices to show that G is onto. This is equivalent to show that the dual map $G^* : (\mathbb{R}^V)^* \to (\mathbb{R}^E)^*$ is injective. Let us identify the dual space $(\mathbb{R}^X)^*$ with $\mathbb{R}^X$ by the standard basis (X is a finite set). Then the dual map $G^* : \mathbb{R}^V \to \mathbb{R}^E$ sends $w \in \mathbb{R}^V$ to $G^*(w)$ so that

$$G^*(w)(e) = w(v_1) + w(v_2)$$

where v_1, v_2 are the end points of e. Now the injectivity of G^* is obvious. For if $G^*(w) = 0$, then $w(u) = -w(v)$ whenever u, v are end points of an edge. Take a triangle σ with three vertices v_1, v_2, v_3. Then

$$w(v_1) = -w(v_2) = w(v_3) = -w(v_1).$$

Thus $w(v_i) \equiv 0$ for $i = 1, 2, 3$. This shows $w = 0$. □

8.3 The work of Rivin and Leibon on Delaunay Polyhedral Surfaces

In the work of [Ri] and [Le], Rivin and Leibon not only established the rigidity theorem of Euclidean and hyperbolic polyhedral metrics on (S,T), they also showed that the space of all ψ_0 curvatures of all Delaubay polyhedral metrics on (S,T) is a convex polytope. We will give a new proof of their theorems in this section.

Theorem 8.6 ([Ri],[Le]). *The space of all ψ_0 curvatures of Delaunay Euclidean or hyperbolic polyhedral metrics on (S,T) is a convex polytope on $\mathbb{R}^E$.*

Proof. We will give a proof for hyperbolic metrics. Similar argument also works for the Euclidean case. See also [Lu1].

Let X be the subset of $P_{\mathbb{H}^2}(S,T) \subset \mathbb{R}^E$ consisting of Delaunay metrics. Let $\Omega = \Psi_0(X)$ be the image of Delaunay metrics under the ψ_0 curvature map $\Psi_0 : P_{\mathbb{H}^2}(S,T) \to \mathbb{R}^E$. We claim that Ω is a convex polytope in $\mathbb{R}^E$ defined by an explicit set of linear inequalities. To this end, we will find the boundary points of Ω. By Leibon's rigidity theorem, Ψ_0 is an embedding. Thus, it suffices to analyze the image under Ψ_0 of the boundary of X in $[0,\infty]^E$. By definition, the set $X \subset [0,\infty]^E$ is given by the following set of inequalities:

$$z(e) < \infty, \tag{8.1}$$

$$z(e) > 0, \tag{8.2}$$

$$z(e_i) + z(e_j) > z(e_k), \tag{8.3}$$

where e_i, e_j, e_k form the boundary of a triangle, and

$$\Psi_0(z) \geqslant 0. \tag{8.4}$$

Note the last inequality (8.4) is a linear constrain for Ω. Thus, it suffices to understand the cases (8.1),(8.2),(8.3). To this end, take a sequence of metrics $l^{(m)} \in X$ so that $\lim_{m\to\infty} l^{(m)}(e)$ exists in $[0,\infty]$ for all $e \in E$ and one of the inequalities (8.1),(8.2) or (8.3) becomes an equality as m tends to ∞. We will analyze the limit points of $\Psi_0(l^{(m)})$. By taking a subsequence if necessary, we may assume that for each corner α in (S,T), the angle of α in the metric $l^{(m)}$, denoted by $\alpha^{(m)}$, converges in $[0,\pi]$ as m tends to infinity.

There are now three cases.

Case 1. There exists $e_1 \in E$ so that $\lim_{m\to\infty} l^{(m)}(e_1) = +\infty$. Let $\triangle$ be a triangle adjacent to e_1 and have edges e_1,e_2,e_3. Let $l_i = \lim_{m\to\infty} l^{(m)}(e_i) \in [0,\infty]$ and α_i be the corner of $\triangle$ facing the edge e_i.

Lemma 8.7. *There exists an index $i\in\{2,3\}$ so that $l_i = \lim_{m\to\infty} l^{(m)}(e_i) = \infty$ and $\lim_{m\to\infty}$ $\alpha_j^{(m)} = 0$, where $\{i,j\} = \{2,3\}$.*

Proof. By the triangle inequality that $l_2 + l_3 \geqslant l_1 = \infty$, we see one of l_2 or $l_3 = \infty$. Now if one of l_2 or l_3 is finite, say $l_3 \in \mathbb{R}$. Then $i = 2$ and $j = 3$. Indeed, by the cosine law,

$$\cos\alpha_3^{(m)} = \frac{-\cosh l^{(m)}(e_3) + \cosh l^{(m)}(e_2)\cosh l^{(m)}(e_1)}{\sinh l^{(m)}(e_2)\sinh l^{(m)}(e_1)} \tag{8.5}$$

and the assumption that $l_1 = l_2 = \infty, l_3 < +\infty$, it follows that

$$\lim_{m\to\infty} \cos(\alpha_3^{(m)}) = 1,$$

i.e.,

$$\lim_{m\to\infty} \alpha_3^{(m)} = 0.$$

In the remaining case that $l_2 = l_3 = \infty$, due to the sum of the two numbers $l_2 - l_3 - l_1$ and $l_3 - l_2 - l_1$ is $-\infty$, we may assume that $l_3 - l_2 - l_1 = -\infty$, i.e.,

$$\lim_{m\to\infty} \frac{\cosh l^{(m)}(e_3)}{\sinh l^{(m)}(e_2)\sinh l^{(m)}(e_1)} = 0.$$

Now use (8.5) again, we conclude that $\lim_{m\to\infty} \alpha_3^{(m)} = 0$. □

Using Lemma 8.7, we find an edge say e_2 of $\triangle$ and the corner α of $\triangle$ adjacent to e_1,e_2 so that $\lim_m \alpha^{(m)} = 0, \lim_m l^{(m)}(e_2) = \infty$.

Let $\triangle^1$ be the triangle adjacent to e_2. By the same argument applied to $\triangle^2$ with respect to e_2 where $\lim_m l^{(m)}(e_2) = \infty$, we find a corner β of $\triangle^2$ adjacent to e_2 and another edge e_2' of $\triangle^2$ so that $\lim_m \beta^{(m)} = 0$ and $\lim_m l^{(m)}(e_2') = +\infty$. Repeating this procedure, we obtain a cycle of edges $\{\varepsilon_1,\cdots,\varepsilon_k\}$ so that ε_i and ε_{i+1} lie in a triangle σ_i so that (i) $\lim_m l^{(m)}(\varepsilon_i) = +\infty$, (ii) $\lim_m \alpha_i^{(m)} = 0$ where α_i is the corner of σ_i adjacent to ε_i and ε_{i+1} and (iii) $\varepsilon_{k+1} = \varepsilon_1$.

Let β_i and γ_i be the corners of σ_i facing ε_i and ε_{i+1} respectively. Then by definition, in the metric $l^{(m)}$,

$$\begin{aligned}\sum_{i=1}^{k} \psi_0(\varepsilon_i) &= \sum_{i=1}^{k}\Big(\frac{1}{2}(\alpha_i^{(m)} + \beta_i^{(m)} - \gamma_i^{(m)}) + \frac{1}{2}(\alpha_{i+1}^{(m)} + \gamma_{i+1}^{(m)} - \beta_{i+1}^{(m)})\Big) \\ &= \sum_{i=1}^{k} \alpha_i^{(m)}.\end{aligned} \tag{8.6}$$

Thus, we see first of all that the right-hand-side of (8.6) is positive due to $\alpha_i^{(m)} > 0$. Furthermore, by condition (ii),

$$\lim_m \sum_{i=1}^k \psi_0(\varepsilon_i) = 0.$$

As a consequence, we showed that $\Omega \subset \{z \in \mathbb{R}^E | \sum_{i=1}^k z(\varepsilon_i) > 0\}$ for any edge cycle $\{\varepsilon_1, \cdots, \varepsilon_k\}$ and $\lim_{m\to\infty} \Psi_0(l^{(m)})$ lies in the hyperplane $\sum_{i=1}^k z(\varepsilon_i) = 0$.

Case 2. Suppose $\lim_m l^{(m)}(e) \in [0,\infty)$ for all $e \in E$ and there exists $e' \in E$ so that $\lim_m l^{(m)}(e) = 0$. Let $I = \{e \in E | \lim_m l^{(m)}(e) = 0\}$. Let F_I be the set of all triangles having an edge in I. If $\triangle \in F_I$, there are two possibilities, due to the triangle inequalities, either (i) all three edges of $\triangle$ are in I or (ii) exactly one edge is in I.

Let α, β, γ be three corners in $\triangle$ so that α faces an edge in I. In case (i), $\lim_m(\alpha^{(m)} + \beta^{(m)} + \gamma^{(m)}) = \pi$ and in case (ii) $\lim_m(\alpha^{(m)}, \beta^{(m)} + \gamma^{(m)}) = (0, \pi)$.

Now consider the sum $\sum_{e\in I} \psi_0(e)$ and break it according to the case (i) or (ii) triangles

$$\sum_{e\in I} \psi_0(e) = \sum_{\sigma\in F_I(i)} \frac{1}{2}(\alpha^{(m)} + \beta^{(m)} + \gamma^{(m)}) + \sum_{\sigma\in F_I(ii)} \frac{1}{2}(\beta^{(m)} + \gamma^{(m)} - \alpha^{(m)})$$

where the sum is over all triangles σ with inner angles $\alpha^{(m)}, \beta^{(m)}, \gamma^{(m)}$, and $F_I(i), F_I(ii)$ denote the sets of triangles satisfying cases (i) and (ii) respectively.

By the Gauss-Bonnet theorem, $\alpha^{(m)} + \beta^{(m)} + \gamma^{(m)} < \pi$. Thus

$$\sum_{e\in I} \psi_0(e) < \frac{1}{2}\pi|F_I|$$

for any $I \subset E$ so that no triangle in T has exactly two edges in I. Furthermore, $\lim_m \sum_{e\in I} \psi_0(e) = \frac{1}{2}\pi|F_I|$. It follows that $\Omega \subset \{z \in \mathbb{R}^E | \sum_{e\in I} z(e) < \frac{\pi}{2}|F_I|\}$ and $\mathrm{Lim}_m \Psi_0(l^{(m)})$ lies in the hyper-surface $\sum_{e\in I} z(e) = \frac{\pi}{2}|F_I|$.

Case 3. For all $e \in E$, $\lim_m l^{(m)}(e) \in \mathbb{R}_{>0}$ and there exists a triangle $\triangle$ with edges e_1, e_2, e_3 so that

$$\lim_m l^{(m)}(e_1) = \lim_m l^{(m)}(e_2) + \lim_m l^{(m)}(e_3). \tag{8.7}$$

We may assume after a different selection that $\lim_m l^{(m)}(e_1)$ is the maximal number among all triangles in the triangulation satisfying (8.7). We claim that Delaunay condition implies this case does not occur. Indeed, let α_i be the corner in $\triangle$ facing the edge e_i. Due to $\lim_m l^{(m)}(e_i) \in \mathbb{R}_{>0}$ and (8.7), the triangle $\triangle$ in the metric $l^{(m)}$ degenerates to a geodesic line segment. It follows that $\lim_m \alpha_1^{(m)} = \pi, \lim_m \alpha_2^{(m)} = \lim_m \alpha_3^{(m)} = 0$. Thus $\lim_m \frac{1}{2}(\alpha_2^{(m)} + \alpha_3^{(m)} - \alpha_1^{(m)}) = -\frac{\pi}{2}$. Let $\triangle^1$ be the triangle adjacent to $\triangle$ along the edge e_1. Let the edges of $\triangle^1$ be e_1, e_4, e_5 and the corner of $\triangle^1$ be $\beta_1, \beta_2, \beta_3$ so that β_1 faces e_1. see Fig. 8.5. Then by the Delaunay condition,

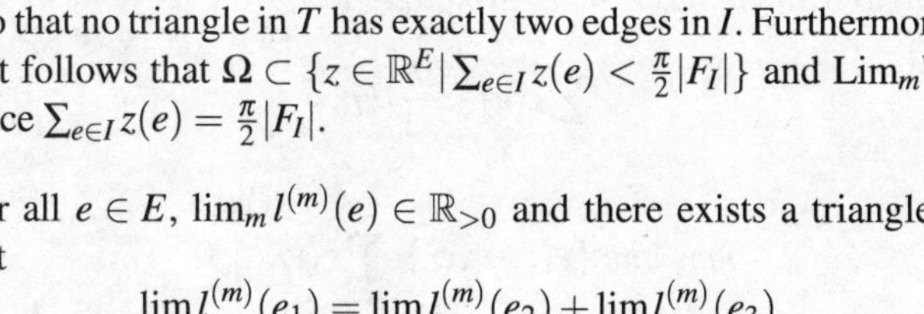

Fig. 8.5.

$$\lim_m \psi_0(e_1) = \lim_m \left(\frac{1}{2}(\alpha_2^{(m)} + \alpha_3^{(m)} - \alpha_1^{(m)}) + \frac{1}{2}(\beta_2^{(m)} + \beta_3^{(m)} - \beta_1^{(m)}) \right) \geqslant 0.$$

But $\lim_m(\alpha_2^{(m)} + \alpha_3^{(m)} - \alpha_1^{(m)}) = -\pi$ and $\beta_2^{(m)} + \beta_3^{(m)} - \beta_1^{(m)} < \pi$. It follows that $\lim_m \beta_1^{(m)} = 0$. Thus, the triangle Δ^1 in the metrics $l^{(m)}$ degenerates. Due to the assumption that $\lim_m l^{(m)}(e) \in \mathbb{R}_{>0}$ for all $e \in E$, it follows that Δ^1 has to be a triangle satisfying (8.7) at its three edges.

Since $\lim_m \beta_1^{(m)} = 0$, we must have

$$\lim_m l^{(m)}(e_5) = \lim_m l^{(m)}(e_1) + \lim_m l^{(m)}(e_4)$$

or

$$\lim_m l^{(m)}(e_4) = \lim_m l^{(m)}(e_1) + \lim_m l^{(m)}(e_5).$$

In both cases, we obtain $\max(\lim_m l^{(m)}(e_4), \lim_m l^{(m)}(e_5)) > \lim_m l^{(m)}(e_1)$. This contradicts the choice of e_1. □

As a consequence, we obtain

Corollary 8.8 (Leibon). *The set of ψ_0 curvature of Delaunay hyperbolic polyhedral metrics on (S,T) is the following convex polytope*

$$\{z \in \mathbb{R}^E_{\geqslant 0} | \text{so that (8.8) and (8.9) hold}\}$$

for $I \subset E$ so that no triangle has exactly two edges in I,

$$\sum_{e \in I} z(e) < \frac{\pi}{2}|F_I|, \tag{8.8}$$

$$\text{new line } \{\varepsilon_1, \cdots, \varepsilon_k\}, \sum_{i=1}^{k} z(\varepsilon_i) > 0. \tag{8.9}$$

Note that (8.8), (8.9) and $z(e) \geqslant 0$ are linear programming problems associated to the triangulation T. The corollary implies the fact that a topological triangulation T of a surface supporting a Delaunay hyperbolic metric with prescribed curvature is a combinatorial property of T. This combinatorial property of T can be checked algorithmically.

9

A Brief Introduction to Teichmüller Space

9.1 Introduction

For simplicity, we consider the compact connected orientable surface of negative Euler characteristic in this chapter. Let $\Sigma_{g,n}$ be a surface of genus $g \geqslant 0$ with $n \geqslant 0$ boundary components. Then its Euler characteristic $\chi(\Sigma_{g,n})$ is $2-2g-n$. Thus negative Euler characteristic means that $(g,n) \neq (0.0),(1,0),(0,2)$ and $(0,1)$.

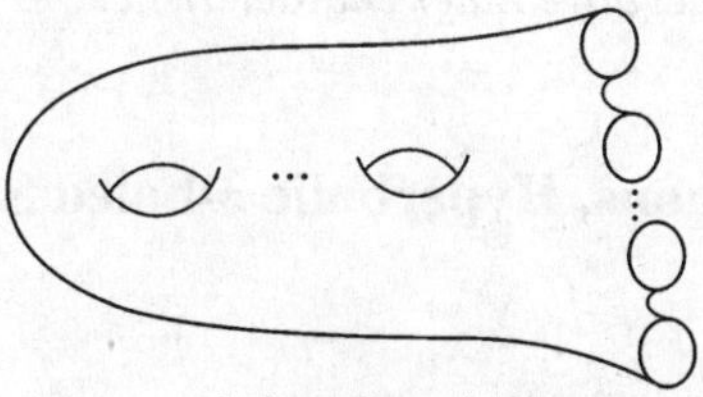

g holes, n boundarics

Fig. 9.1.

We will see below that each $\Sigma_{g,n}$ with $\chi(\Sigma_{g,n}) < 0$ supports a hyperbolic metric with geodesic boundary. The question naturally arising is then how many different hyperbolic metrics one can find on $\Sigma_{g,n}$. This is the moduli space problem first studied by B.Riemann. To be more precise, Riemann's moduli space of a surface Σ, denoted by Mod(Σ) is the set of isometry classes of hyperbolic metrics with geodesic boundary on the surface. The moduli space Mod(Σ) has complicated topology and is difficult to study. Teichmüller's major contribution was that he introduced a finer equivalence relationship between hyperbolic metrics on a surface. In this way, the study of the moduli space of a surface is divided into two parts.

Definition 9.1. *Two hyperbolic metrics d,d' on a surface Σ are (Teichmüller) equivalent if there is an isometry between (Σ,d) and (Σ,d') so that the isometry is homotopic to the identity. The set of all equivalence classes of all hyperbolic metrics with geodesic boundary on Σ is called the Teichmüller space of the surface, denoted by* Teich(Σ).

By definition, the Riemann's moduli space Mod(Σ) is the quotient space of the Teichmüller space Teich(Σ) acted by a group $\Gamma(\Sigma)$. Here $\Gamma(\Sigma)$ is the group of homotopy classes of self-homeomorphisms of the surface Σ. The group $\Gamma(\Sigma)$ is called the mapping class group of the surface.

One of the best ways to illustrate the situation is to consider the space of triangles.

Example 9.2. *A marked triangle is a triangle whose three edges are labelled by 1,2 and 3 respectively. Two marked Euclidean triangles are equivalent if there is an isometry between them preserving the labelling. Then the (Teichmüller) space of all equivalence classes of marked Euclidean triangles is*

$$E_l(2) = \{(l_1, l_2, l_3) | l_i + l_j > l_k, \quad \{i, j, k\} = \{1, 2, 3\}\}$$

parameterized by the edge lengths. On the other hand, the space of all isometry classes of Euclidean triangles is $E_l(2)/S_3$ where S_3 is the permutation group of three elements $\{1,2,3\}$ acting naturally on $E_l(2)$. The space $E_l(2)/S_3$ is the counterpart of Riemann's moduli space. The group S_3 is the analogy of the mapping class group $\Gamma(\Sigma)$. Due to the fixed points of S_3 action, the space $E_l(2)/S_3$ contains singularities (an orbifold).

As we will see, the Teichmüller space Teich(Σ) (in the natural topology) is always homeomorphic to a cell much like $E_l(2)$.

We end this introduction by mentioning the important uniformization theorem.

Theorem 9.3 (Poincaré-Köbe). *If M^2 is a connected surface without boundary and ds is a Riemannian metric on M^2, then there exists a complete constant curvature 0,1, or -1 metric on M^2 which is conformal equivalent to ds. The constant curvature metric is unique unless M^2 has non-negative Euler characteristic.*

9.2 Hyperbolic Hexagons, Hyperbolic 3-holed Spheres and the Cosine law

We begin with the following lemma.

Lemma 9.4. *For any $l_1, l_2, l_3 \in \mathbb{R}_{>0}$, there exists a right angled hyperbolic hexagon whose three non-pairwise adjacent edge lengths are l_1, l_2, l_3. Furthermore, the hexagon is unique up to isometry.*

Proof. We use the following fact to prove the lemma. Given a geodesic l in the upper half-plane $\mathbb{H}^2$ and $d > 0$, the set $\{x \in \mathbb{H}^2 | d(x,l) = d\}$ consists of two circular arcs α, β which intersect l at the end points. Indeed, we may assume after applying an isometry that l is the **y**-axis. Then the isometry $f(z) = \lambda z$ for $\lambda > 0$ leaves l invariant. It follows that $d_{\mathbb{H}}(z,l) = d_{\mathbb{H}}(\lambda z, l)$. This shows that α, β are two rays from the origin. see Fig. 9.2. Now to prove Lemma 9.4, take two disjoint geodesics a, b of distance l_1 apart in the disk

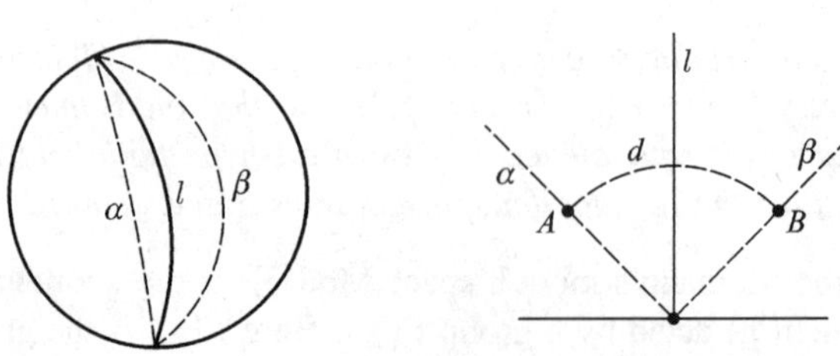

Fig. 9.2.

model of the hyperbolic plane $\mathbb{H}^2$. Consider two circular arcs a', b' in the common region bounded by a, b so that a' (or b') consists of points of distance l_2 (or l_3) to a (or b).

Finally, draw a geodesic c tangent to both a', b'. (There are two choices of c which are related by the hyperbolic reflection about the common perpendicular of a, b). see Fig. 9.3.

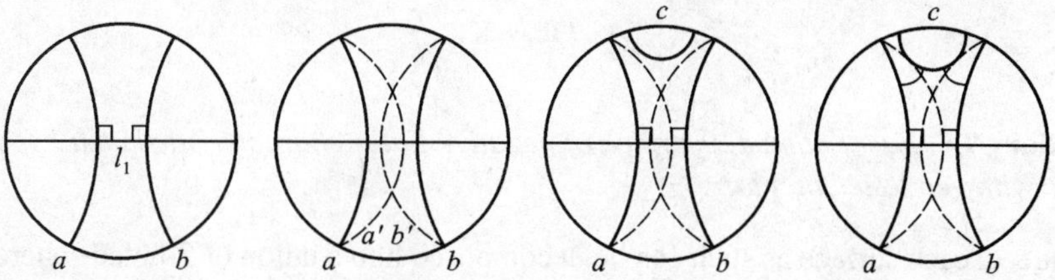

Fig. 9.3.

Now draw the common perpendicular arcs between all three pairs $\{a,b\}, \{b,c\}$ and $\{c,a\}$. We obtain a right angled hyperbolic haxagon of the prescribed pairwise non-adjacent edge lengths l_1, l_2, l_3. The uniqueness follows from the construction. □

As a consequence, we have

Corollary 9.5. *Given any $l_1, l_2, l_3 \in \mathbb{R}_{>0}$, there exists a unique hyperbolic metric with geodesic boundary on the 3-holed sphere $\Sigma_{0,3}$ so that the lengths of the boundary components are l_1, l_2, l_3.*

Proof. Take two right-angle hyperbolic right-angled hexagons whose pairwise non-adjacent edges e_1, e_2, e_3 have lengths $l_1/2, l_2/2, l_3/2$. Now isometrically identify the other three edges ($\neq e_i$) in pairs. The result is a hyperbolic metric on the 3-holed sphere whose boundary are geodesics of lengths l_1, l_2, l_3. see Fig. 9.4.

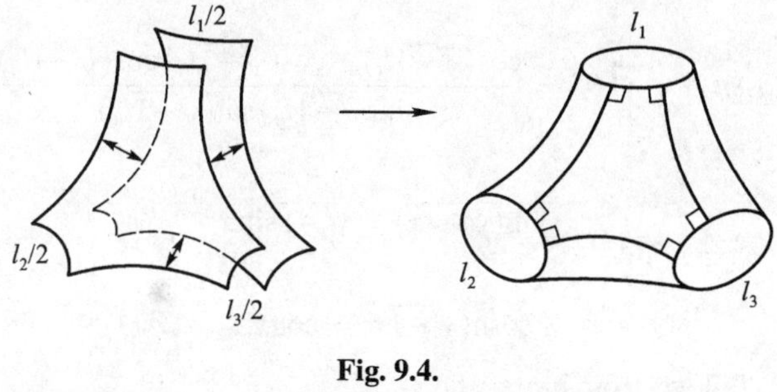

Fig. 9.4.

To see the uniqueness, if d is a hyperbolic metric on $\Sigma_{0,3}$ with geodesic boundary, draw the shortest geodesic arc in $\Sigma_{0,3}$ between each pair of boundary components. Say $\alpha_1, \alpha_2, \alpha_3$ are the arcs. Then α_i is orthogonal to the boundary and $\alpha_i \cap \alpha_j = \emptyset$ for $i \neq j$. Indeed, if otherwise that $\alpha_i \cap \alpha_j \neq \emptyset$, there would exist a hyperbolic triangle whose inner angles are $\pi/2, \pi/2$, and a joint together (see Fig. 9.5). This will contradict Gauss-Bonnet theorem. Now cut the surface $\Sigma_{0,3}$ open along $\alpha_1 \sqcup \alpha_2 \sqcup \alpha_3$ to obtain two right-angle hyperbolic hexagons P, P'. Since P, P' have the same lengths at three pairwise non-adjacent edges, there are isometric by the uniqueness part of Lemma 9.4. It follows that the metric on d is obtained by the metric double construction. □

As a consequence of Corollary 9.5, we have

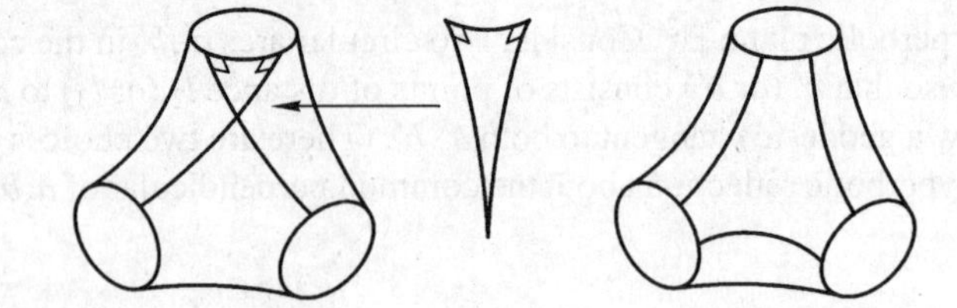

Fig. 9.5.

Corollary 9.6. *Every compact surface of negative Euler characteristic has a hyperbolic metric with geodesic boundary.*

Indeed, each surface as such can be decomposed into a union of 3-holed spheres by a set of disjoint simple closed curves. Now realize each of the 3-holed sphere by a hyperbolic 3-holed sphere with geodesic boundary so that each boundary component has length 1. The isometric gluing of these hyperbolic 3-holed spheres produces a hyperbolic metric on the original surface.

Given a hyperbolic right-angled hexagon with three non-pairwise adjacent edge lengths l_1, l_2, l_3 and opposite edge lengths $\theta_1, \theta_2, \theta_3$, the cosine law says,

$$\cosh\theta_i = \frac{\cosh l_i + \cosh l_j \cosh l_k}{\sinh l_j \sinh l_k}$$

where $\{i,j,k\} = \{1,2,3\}$. We have

(1) $\dfrac{\partial\theta_i}{\partial l_i} = \dfrac{\sinh\theta_i}{A} > 0$,

(2) $\dfrac{\partial\theta_i}{\partial l_j} = \dfrac{\partial\theta_i}{\partial l_i}\cosh\theta_k$.

Identity (1) follows from a direct derivative calculation. To see (2), we rewrite the cosine law above as

$$\cos(\pi - \sqrt{-1}\theta_i) = \frac{\cos(\sqrt{-1}l_i) + \cos(\sqrt{-1}l_j)\cos(\sqrt{-1}l_k)}{\sin(\sqrt{-1}l_j)\sin(\sqrt{-1}l_k)}$$

due to

$$\sinh(\sqrt{-1}x) = \sqrt{-1}\sin x$$

and

$$\cosh(\sqrt{-1}x) = \cos x.$$

Then identity (2) follows from proposifion 4.2.

By the tangent law, Lemma 7.4,

(3) $\tanh\dfrac{l_i}{2}\cosh r_i$ is independent of the index i.

Theorem 9.7. *For a hyperbolic right-angled hexagon of three non-pairwise adjacent edge lengths l_i and opposite edge lengths $r_j + r_k$, $\{i,j,k\} = \{1,2,3\}$, the differential 1-form*

$$\eta_h = \sum_{i=1}^{3} \frac{\int^{l_i}\coth^h(t/2)dt}{\cosh^{h+1} r_i} dr_i$$

is closed. Its integration $\int^u \eta_h$ is locally strictly concave in u where $u_i = \int_1^{r_i}\cosh^{h-1} t dt$.

Proof. The proof of closeness of the 1-forms is just a specialization of Theorem 7.6 by taking care of the curvature factors. Indeed, we take $y_i = \pi - \sqrt{-1}l_i$ and $x_i = \sqrt{-1}\theta_i$ in Theorem 7.6.

To prove the convexity of the function $F(u) = \int^u \eta_h$, we use Lemma 7.4 for $y_i = \pi - \sqrt{-1}l_i$ and $x_i = \sqrt{-1}\theta_i$. The tangent law in Lemma 7.4 shows $\coth \frac{l_i}{2} / \cosh r_i$ is independent of the indices and $\frac{1}{\cosh r_i}\frac{\partial l_i}{\partial r_j}$ is symmetric in i, j. Now the Hessian $H = [H_{st}]$ of the function $F(u) = \int^u \eta_h$ where $f'(l) = \coth^h(l/2)$ and $g'(r) = \cosh^{-h-1} r$, is

$$
\begin{aligned}
H_{st} &= \frac{f'(l_s)}{g'(r_t)}\frac{\partial l_s}{\partial r_t} \\
&= \coth^h(l_s/2)\cosh^{h+1} r_t \frac{\partial l_s}{\partial r_t} \\
&= \left(\frac{\coth(l_s/2)}{\cosh r_s}\right)^h (\cosh r_t \cosh r_s)^{h+1}\left(\frac{1}{\cosh r_s}\frac{\partial l_s}{\partial r_t}\right) \\
&= \left(\frac{\coth(l_s/2)}{\cosh r_s}\right)^h (\cosh r_t \cosh r_s)^{h+1}\left(\frac{1}{\cosh r_s} a_{st}\right).
\end{aligned}
$$

By the tangent law, $\coth(l_s/2)/\cosh r_s > 0$ is independent of the indices. Above identity shows that the Hessian H is congruent to the matrix $[a_{st}]$ where for $i \neq j$,

$$
\begin{aligned}
a_{ij} &= \frac{1}{\cosh r_i}\frac{\partial l_i}{\partial r_j} \\
&= \frac{1}{\cosh r_i}\left(\frac{\partial l_i}{\partial \theta_i} + \frac{\partial l_i}{\partial \theta_k}\right) \\
&= -\frac{1}{\cosh r_i}\frac{\partial l_i}{\partial \theta_i}(\cosh l_j - 1).
\end{aligned}
$$

On the other hand, we have

$$
\begin{aligned}
a_{ii} &= \frac{1}{\cosh r_i}\frac{\partial l_i}{\partial r_i} \\
&= -\frac{1}{\cosh r_i}\frac{\partial l_i}{\partial \theta_i}(\cosh l_k + \cosh l_j).
\end{aligned}
$$

But $\frac{\partial l_i}{\partial \theta_i} > 0$, this shows that $a_{st} < 0$ for all s, t and $|a_{ii}| > |a_{ij}| + |a_{ik}|$ for all i. Thus the matrix $-[a_{st}]$ is diagonally dominated. It follows that the Hessian matrix H is negative definite. Thus the integration $\int^u \eta_h$ is locally strictly concave. □

9.3 Ideal Triangulation of Surfaces and the Length Coordinate of the Teichmüller Spaces

Recall that a closed triangulated surface (X, T^*) is obtained as follows. Take a finite disjoint union of Euclidean triangles and identify all edges of triangles in pairs by homeomorphisms. The quotient is a closed surface X together with a triangulation T^*. The cells in T^* are the quotients of the vertices, edges and triangles in the unidentified space.

For a compact surface with non-empty boundary, the most efficient way of decomposing it is the *ideal triangulation.* Here is the definition. Take a closed triangulated surface

(X, T^*). Let $N(V)$ be a small open regular neighborhood of the union of all vertices V. Then $S = X - N(V)$ is a compact surface with $|V|$ many boundary components. The set $T = \{\sigma \cap S | \sigma \in T^*\}$ is called an *ideal triangulation* of the surface S. The set of all edges (respectively 2-cells, or hexagons) in T is $\{e \cap S | e \text{ an edge in } T^*\}$ (respectively $\{\sigma \cap S | \sigma \text{ a triangle in } T^* \}$). Each hexagon in T has three edges in $T^{(1)}$ (even though two of the three edges may be the same in $T^{(1)}$). The intersection of the hexagon with the boundary ∂S consists of three arcs called *B-arcs*. A B-arc in a hexagon $\sigma^2 \cap S$ corresponds to a corner in the triangle σ^2 in T^*. The concepts of a B-arc *facing* (or *adjacent to*) an edge e in T are defined as before. For instance, a B-arc is facing to exactly one edge and is adjacent to two edges.

Another way of introducing the ideal triangulation is as follows. A *colored hexagon* is a hexagon with three non-pairwise adjacent edges labelled by red and the opposite edges labelled by black. Take a finite disjoint union of colored hexagons and identify all red edges in pairs by homeomorphisms. The quotient is a collection of compact surfaces with non-empty boundary together with ideal triangulations. The 2-cells in the ideal triangulation are quotients of the hexagons. The quotients of red-edges (respectively black-edges) are the edges (respectively B-arcs) in the ideal triangulation.

The following result is simple and its proof will be left as a homework.

Lemma 9.8. *Each compact surface Σ with $\partial\Sigma \neq \emptyset$ and negative Euler characteristic admits an ideal triangulation. Furthermore, if $\Sigma = \Sigma_{g,n}$, then the number of edges in the ideal triangulation is $6g - 6 + 3n$.*

We now recall the length coordinate of the Teichmüller space of a surface with an ideal triangulation. The situation is much like the polyhedral metrics on closed triangulated surfaces. Given an ideally triangulated surface (S, T) with E as the set of edges and a function $l : E \to \mathbb{R}_{>0}$, one uses Lemma 9.4 to produces a hyperbolic metric, denoted by d_l, with totally geodesic boundary on S. This metric is constructed by making each 2-cell H in $T^{(2)}$ with edges e_i, e_j, e_k a colored hyperbolic right-angled hexagon with red edge lengths $l(e_i), l(e_j), l(e_k)$ and glue isometrically these hexagons along their red edges. This results a hyperbolic metric d_l on the surface. Thus we have a map from $\mathbb{R}^E_{>0}$ to Teich(S). We claim that this is a bijection. As a consequence, we call $l \in \mathbb{R}^E_{>0}$ the length parameter of the Teichmüller space (see Fig. 9.6).

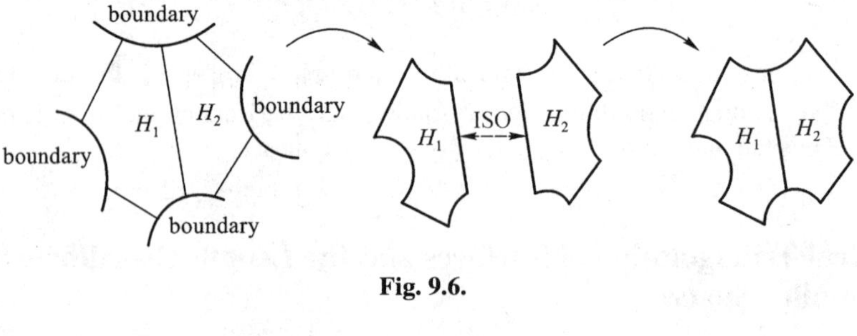

Fig. 9.6.

To see the bijectivity, we need to show that
(1) If $l \neq l' : E \to \mathbb{R}_{>0}$, then the associated metrics d_l and $d_{l'}$ are not equivalent.
(2) If d is a hyperbolic metric on Σ with geodesic boundary, then $d \cong d_l$ for some $l : E \to \mathbb{R}_{>0}$.

To see (1), since $l \neq l'$, there exists an edge $e \in E$ so that $l(e) \neq l'(e)$. This shows that $d_l \neq d_{l'}$. To see (2), we use the same method of proof appeared in Corollary 9.5. Suppose d is a hyperbolic metric on (S,T). For each edge $e \in T^{(1)}$, let e^* be the shortest arc homotopic to e relative to the boundary ∂S. Since e^* is the shortest, the geodesic arc e^* is perpendicular to the boundary of S and e^* is an embedded arc. Furthermore, if $e_1 \neq e_2$ are different arcs in E, then their geodesic representatives e_1^* and e_2^* are disjoint by the Gauss-Bonnet lemma, as shown in Fig. 9.7. In fact, it is a general property of closed geodesics on hyperbolic surfaces. If a and b are two disjoint, non-homotopic, and homotopically essential arcs or loops in a hyperbolic surface, then their geodesic representatives are disjoint. See for instance, Casson-Bleiler's book [CB] for a proof.

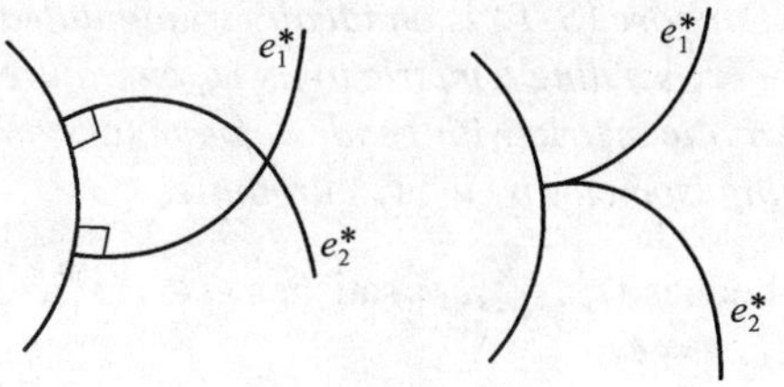

Fig. 9.7.

Thus, the set $T^{*(1)} = \{e^* \mid e \in T^{(1)}\}$ produces a decomposition of S into $\mathbb{H}^2$ right angled hexagons. Let $l : E \to \mathbb{R}_{>0}$ where $l(e)$ is the length of e^*. Then by the construction, we have $d \cong d_l$.

To summarize, we have proved the following proposition.

Proposition 9.9. *Suppose (S,T) is an ideal triangulated surface. Then the Teichmüller space* Teich(S) *can be parameterized by the length functions $l : E \to \mathbb{R}_{>0}$ so that*

$$\text{Teich}(S) \cong \mathbb{R}^E_{>0}.$$

Clearly, there is a strong similarity between Proposition 9.9 and the space of all hyperbolic polyhedral metrics on a closed triangulated surface, denoted by $P(\Sigma,T)$. By definition, in the length parameterization, $P(\Sigma,T)$ is an open convex polyhedral subset of $\mathbb{R}^E_{>0}$ defined by the triangular inequalities. Furthermore, the notion of B-arcs corresponds to the concept of corner for closed surfaces. This prompts us to study the relationship between the length coordinate and the length of the B-arcs.

9.4 New Coordinates for the Teichmüller Space

Suppose a right-angled hyperbolic hexagon has edge lengths l_1, l_2, l_3 at three pairwise non-adjacent edges and the length of the opposite edges are $\theta_1, \theta_2, \theta_3$. Let us recall that by Theorem 9.7, for all $h \in \mathbb{R}$,

$$\omega_h = \sum_{i=1}^{3} \left(\int_1^{l_i} \tanh^{h+1} \frac{t}{2} dt \right) d \left(\int_0^{r_i} \cosh^h t dt \right) = \sum_{i=1}^{3} \left(\int_1^{l_i} \tanh^{h+1} \frac{t}{2} dt \right) dv_i$$

is closed. Furthermore, $F(v) = \int_a^v \omega_h$ is strictly concave in v. Taking Legendre transform of F, we obtain a closed 1-form $\eta_h = \Sigma_{i=1}^3 (\int_0^{r_i} \cosh^h t dt) du_i$, where $u_i = \int_1^{l_i} \tanh^{h+1} t dt$ and its integration $g(u) = \int_0^u \eta_h$ is strictly concave in u, so that

$$\frac{\partial g}{\partial u_i} = \int_0^{r_i} \cosh^h t dt.$$

Suppose d is hyperbolic metric on (S,T) with $\partial S \neq \emptyset$ and T is an ideal triangulation, define its ψ_h curvature $\psi_h : E \to \mathbb{R}$ by

$$\psi_h(e) = \int_0^{\frac{a+b-c}{2}} \cosh^h t dt + \int_0^{\frac{a'+b'-c'}{2}} \cosh^h t dt$$

where a,b,a',b' are the lengths of B-arcs adjacent to e, and c,c' are the lengths of B-arcs facing e. We will prove the following results in the next chapter.

Theorem 9.10 ([Lu1]). *Suppose (S,T) is an ideally triangulated surface. For any $h \in \mathbb{R}$, the map $\Psi_h : \mathrm{Teich}(S) \to \mathbb{R}^E$ sending a metric to its ψ_h curvature is a smooth embedding. In particular, each hyperbolic metric with geodesic boundary on (S,T) is determined up to triangulation preserving isometry by its ψ_h curvature.*

Recall that a cycle of edges $\{e_1,\cdots,e_k\}$ satisfies $e_i \in T^{(1)}$ and for all indices i, e_i, e_{i+1} lie in a hexagon where $e_{k+1} = e_1$.

Theorem 9.11 ([Lu2]). *Let $h \geqslant 0$. For an ideal triangulated surface (S,T), $\Psi_h(\mathrm{Teich}(S)) = \{z \in \mathbb{R}^E \mid$ for each edge cycle $\{e_1,\cdots,e_m\}$, $\sum_{i=1}^m z(e_i) > 0\}$. In particular, the image is an open convex polytope.*

In the case $h = 0$, we can draw the following interesting conclusion. Given a compact surface S with boundary components $b_1,\cdots,b_m$ and $k = (k_1,\cdots,k_m) \in \mathbb{R}^m_{>0}$, the *bordered Teichmüler space* $\mathrm{Teich}_b(S,k)$ consists of those classes of hyperbolic metrics on S so that the length of the boundary geodesic b_i is k_i. This space has appeared many times in recent works of Teichmüller spaces ([Mi], etc). In the case $h = 0$, we can recover the length of the boundary from the ψ_0-curvature, like Lemma 8.4, as follows. Each boundary component b_i of S is associated with an edge cycle $\{e_{in_1},\cdots,e_{in_{m_i}}\}$ where each e_{in_j} has one end point in b_i. Then the length $l(b_i)$ of the boundary b_i is given by $\sum_{j=1}^{m_i} \psi_0(e_{in_j}) = b_i$. Thus, as a corollary, we have

Corollary 9.12 ([Lu3]). *The image of the bordered Teichmüller space $\mathrm{Teich}_b(S,k)$ under the coordinate map Φ_0 is a convex polytope.*

10

Parameterizatios of Teichmüller spaces

For an ideal triangulated surface (S,T), and $h \in \mathbb{R}$, we produce a parameterization Ψ_h of the Teichmüller space of the surface in this chapter. This parameterization is the counterpart of ψ_h curvature for polyhedral metrics.

Let the set of all edges in (S,T) be E. Write $E = \{e_1, \cdots, e_n\}$. If $f : E \to X$ is a function, we use f_i to denote $f(e_i)$.

Let $h \in \mathbb{R}$. Given a hyperbolic metric l on (S,T), the *ψ_h-curvature* of the metric is a map from E to $\mathbb{R}$ defined by

$$\psi_h(e) = \int_0^{\frac{a+b-c}{2}} \cosh^h t dt + \int_0^{\frac{a'+b'-c'}{2}} \cosh^h t dt \qquad (10.1)$$

where a, b, a', b' are the lengths of the B-arcs adjacent to the edge e and c, c' are the lengths of the B-arcs facing the edge e.

We identity the Teichmuller space Teich(S) with $\mathbb{R}^E_{>0}$ by the edge length function $l : E \to \mathbb{R}_{>0}$ and denote $\Psi_h : \mathbb{R}^E_{>0} \to \mathbb{R}^E$ (also Ψ_h : Teich$(S) \to \mathbb{R}$) the map sending a metric l to its ψ_h curvature.

Theorem 10.1 ([Lu1]). *Suppose (S,T) is an ideally triangulated surface. For any $h \in \mathbb{R}$, the map $\Psi_h : T(S) \to \mathbb{R}^E$ is a smooth embedding.*

In particular, the theorem says each hyperbolic metric with geodesic boundary on (S,T) is determined up to isometry by its ψ_h curvature.

Recall that a *cycle of edges* $\{e_{n_1}, \cdots, e_{n_k}\}$ is a subset of E so that for all indices i, $e_{n_i}, e_{n_{i+1}}$ lie in a hexagon where $e_{n_{k+1}} = e_{n_1}$.

Theorem 10.2 ([Lu1]). *Let $h \geqslant 0$. For an ideal triangulated surface (S,T), $\Psi_h(\text{Teich}(S)) = \{z \in \mathbb{R}^E \mid$ for each edge cycle $\{e_{n_1}, \cdots, e_{n_k}\}$, $\sum_{i=1}^k z(e_{n_i}) > 0\}$. In particular, the image is an open convex polytope independent of h.*

Remark 10.3. *Ren Guo [Gu1] has recently proved that for $h < 0$, the image $\Psi_h(\text{Teich}(S))$ is also a convex polytope.*

Remark 10.4. *Whether these new coordinates are related to the quantum Teichmüller space ([FC], [Ka], [BL]) is an interesting question.*

Remark 10.5. *An edge cycle $\{e_{n_1}, \cdots, e_{n_k}\}$ is called fundamental* if each edge appears at most twice. It is proved in [Lu2] that the convex set $\Psi_h(\text{Teich}(S))$ is defined by a finite set of inequalities $\sum_{i=1}^m z(e_{n_i}) > 0$ where $\{e_{n_1}, \cdots, e_{n_k}\}$ are fundamental cycles. Thus, $\Psi_h(\text{Teich}(S))$ is an open convex polytope in $\mathbb{R}^E$.

10.1 A Proof of Theorem 10.1

The proof of Theorem 10.1 is a simple application of the strictly convexity of the energy functions introduced in §9.5. For a colored hyperbolic right-angled hexagon of red edge lengths l_1, l_j, l_k and opposite black edge lengths $\theta_i, \theta_j, \theta_k$ where $\theta_i = r_j + r_k$, $\{i, j, k\} = \{1, 2, 3\}$, the following 1-form

$$\omega = \sum_{i=1}^{3} \int_0^{r_i} \cosh^h t dt d\left(\int_1^{l_i} \tanh^{h+1}(t/2)dt\right)$$

is closed. Indeed, the 1-form obtained by taking the Legendre transformation of the integration obtained in Theorem 9.7. Let $u_i = \int_1^{l_i} \tanh^{h+1}(t/2)dt$ and $u = (u_1, u_2, u_3)$. Then the integration $F(u) = \int_{(1,1,1)}^{u} w$ is strictly concave in $u \in I^3$ where $I = g(\mathbb{R}_{>0})$ for the function $g(x) = \int_1^x \tanh^{h+1}(t/2)dt$.

Given a hyperbolic metric $l : E \to \mathbb{R}_{>0}$ on (S, T), let $u : E \to \mathbb{R}$ be $u(e) = g(l(e))$ $(= \int_1^{l(e)} \tanh^{h+1}(t/2)dt)$. Then the set of all possible values of u forms the open convex cube I^E. Define an energy function $W : I^E \to \mathbb{R}$ by

$$W(u) = \sum_{\{e_a, e_b, e_c\} \in T^{(2)}} F(u_a, u_b, u_c)$$

where $u_i = u(e_i)$ and the sum is over all hexagons with edges e_a, e_b, e_c. By definition, W is smooth and strictly concave in I^E. Furthermore, by the construction of F,

$$\frac{\partial W}{\partial u_i} = \psi_h(e_i),$$

i.e., $\bigtriangledown W = \Psi_h$. By Lemma 5.4 on gradient of strictly convex function, we conclude that the map $\bigtriangledown W : I^E \to \mathbb{R}^E$ is a smooth embedding. This proves Theorem 10.1.

10.2 Degenerations of Hyperbolic Hexagons

A proof of Theorem 10.2 depends on a study of degenerations of hyperbolic hexagons.

Lemma 10.6. *Suppose a hyperbolic right-angled hexagon has three non-pairwise adjacent edge lengths l_1, l_2, l_3 and opposite edge lengths $\theta_1, \theta_2, \theta_3$ so that $\theta_i = r_j + r_k$, $\{i, j, k\} = \{1, 2, 3\}$. Then the following hold.*
(a) $\lim_{\theta_i \to 0} l_j(\theta_1, \theta_2, \theta_3) = \infty$ *for* $j \neq i$.
(b) $\lim_{l_i \to 0} |r_i| = \infty$.
(c) *Suppose a sequence of hexagons satisfy $|r_1|, |r_2|, |r_3|$ are uniformly bounded. Then* $\lim_{l_i \to \infty} \theta_j \theta_k = 0$.

Proof. For (a), we use the cosine law that

$$\begin{aligned}\cosh l_j &= \frac{\cosh\theta_j + \cosh\theta_i \cosh\theta_k}{\sinh\theta_i \sinh\theta_k} \\ &\geqslant \frac{\cosh\theta_i \cosh\theta_k}{\sinh\theta_i \sinh\theta_k} \\ &\geqslant \coth\theta_i.\end{aligned}$$

Since $\lim_{\theta_i \to 0} \coth\theta_i = \infty$, it follows from the above inequality that $\lim_{\theta_i \to 0} l_j = \infty$.

For (b), we use the tangent law that

$$
\begin{aligned}
\tanh^2(l_i/2) &= \frac{\cosh r_j \cosh r_k}{\cosh r_i \cosh(r_i+r_j+r_k)} \\
&= \frac{1}{\cosh r_i(\cosh r_i(\cosh r_i(1+\tanh r_j \tanh r_k)+\sinh r_i(\tanh r_j+\tanh r_k)))} \\
&\geqslant \frac{1}{\cosh r_i(\cosh r_i(1+1)+|\sinh r_i|(1+1))} \\
&\geqslant \frac{1}{4\cosh^2 r_i}.
\end{aligned}
$$

It follows that $\cosh^2 r_i \geqslant \frac{1}{4\tanh^2(l_i/2)}$. Thus part (b) follows.

For part (c), by the assumption that $|r_i|$'s are uniformly bounded, it follows that $\theta_i = r_j + r_k$ are uniformly bounded from above. Now the cosine law says that

$$
\begin{aligned}
\cosh l_i &= \frac{\cosh\theta_i + \cosh\theta_j \cosh\theta_k}{\sinh\theta_j \sinh\theta_k} \\
&\leqslant \frac{C}{\sinh\theta_j \sinh\theta_k}
\end{aligned}
$$

for some constant C. Thus $\sinh\theta_j \sinh\theta_k \leqslant \frac{C}{\cosh l_i}$. Since θ_j and θ_k are uniformly bounded from above, it follows that $\lim_{l_i\to\infty} \theta_j\theta_k = 0$. □

10.3 A Proof of Theorem 10.2.

Let $\Psi_h : \mathbb{R}^E_{>0} \to \mathbb{R}^E$ be the map sending a hyperbolic metric $l \in \mathbb{R}^E_{>0}$ to its ψ_h curvature. Let Ω be the convex set $\{z \in \mathbb{R}^E |$ whenever $e_{i_1}, \cdots, e_{i_k}$ form an edge cycle, $\sum_{j=1}^k z(e_{i_j}) > 0\}$. We will prove the theorem in two steps. In the first step we prove $\Psi_h(\mathbb{R}^E_{>0}) \subset \Omega$. In the second step, we will prove that $\Psi_h(\mathbb{R}^E_{>0})$ is a closed subset of Ω. Since Theorem 10.1 shows that $\Psi_h(\mathbb{R}^E_{>0})$ is open in Ω. It follows from the connectivity of Ω that $\Psi_h(\mathrm{Teich}(S)) = \Omega$.

To establish the first step that $\Psi_h(\mathbb{R}^E_{>0}) \subset \Omega$, we fix a hyperbolic metric $l \in \mathbb{R}^E_{>0}$. Take an edge cycle $\{e_{n_1}, \cdots, e_{n_k}\}$, let a_i be the length of the B-arc in the hexagon H_i containing both e_{n_i} and $e_{n_{i+1}}$ (where $e_{n_{k+1}} = e_{n_1}$) so that this B-arc is adjacent to e_{n_i} and $e_{n_{i+1}}$. Denote the lengths of B-arcs in H_i facing e_{n_i} and $e_{n_{i+1}}$ by b_i and c_i. Then by definition, the contribution to $\sum_{i=1}^m z(e_i)$ from the B-arcs inside H_i is

$$\int_0^{\frac{a_i+b_i-c_i}{2}} \cosh^h t dt + \int_0^{\frac{a_i+c_i-b_i}{2}} \cosh^h t dt > 0.$$

The last inequality follows from the fact that the function $F(t) = \int_0^{t+c} \cosh^h s ds + \int_0^{t-c} \cosh^h s ds$ is a strictly increasing function of t in the interval $[0,\infty]$ and $F(0) = 0$ for any constant c. It follows that the sum $\sum_{i=1}^k z(e_{n_i}) = \sum_{i=1}^k \int_0^{\frac{a_i+b_i-c_i}{2}} \cosh^h t dt + \int_0^{\frac{a_i+c_i-b_i}{2}} \cosh^h t dt > 0$.

To establish the second step that $\Psi_h(\mathbb{R}^E_{>0})$ is closed in Ω, take a sequence $l^{(m)} \in \mathbb{R}^E_{>0}$ so that $\Psi_h(l^{(m)})$ converges to a point $b \in \Omega$. We claim that $l^{(m)}$ contains a subsequence converging to a point $a \in \mathbb{R}^E_{>0}$. By taking subsequence, we may assume that $\lim_{m\to\infty} l^{(m)} = a \in [0,\infty]^E$. It remains to show that for each edge $e \in E$, $a(e) \in (0,\infty)$.

Suppose otherwise that there is an edge $e \in E$ so that $a(e) \in \{0,\infty\}$. We will discuss two cases.

Case 1. $a(e) = 0$ for some $e \in E$, i.e., $\lim_{m\to\infty} l^{(m)}(e) = 0$. Let H, H' be the hexagons containing e and r, r' be the r-coordinates of the B-arcs in H, H' facing e. (Recall that if the lengths of the B-arcs in H are x, y, z so that B-arc of length x is facing the edge e, then the r-coordinate at the the B-arc is $\frac{y+z-x}{2}$.) By Lemma 10.6(b), in the metric $l^{(m)}$, we have $\lim_{m\to\infty} |r| = \lim_{m\to\infty} |r'| = \infty$. Due to the assumption that $b \in \Omega$,

$$z(e) = \int_0^r \cosh^h t dt + \int_0^{r'} \cosh^h t dt$$

is uniformly bounded in the metrics $l^{(m)}$. On the other hand, $h \geqslant 0$ implies that $\int_0^\infty \cosh^h t dt = \infty$. It follows that one of the limits $\lim_{m\to\infty} r$ and $\lim_{m\to\infty} r'$ is ∞ and other one is $-\infty$. Say $\lim_{m\to\infty} r = -\infty$. Let r_1 and r_2 be the r-coordinates of the other two B-arcs in the hexagon H. Since $r_i + r \geqslant 0$ by definition of r-coordinate for $i = 1,2$, we obtain $\lim_{m\to\infty} r_i = \infty$ for $i = 1,2$ in the sequence of the metrics $l^{(m)}$.

In summary, we obtain two rules governing the r-coordinates of B-arcs in hexagons H in the metrics $l^{(m)}$.

Rule I. If a B-arc has r-coordinate converging to $-\infty$, then the other two B-arcs in the same hexagon have r-coordinates converging to ∞.

Rule II. If a and b are two different B-arcs facing an edge so that the r-coordinate of a converges to $\pm\infty$, then the r-coordinate of b converges to $\mp\infty$.

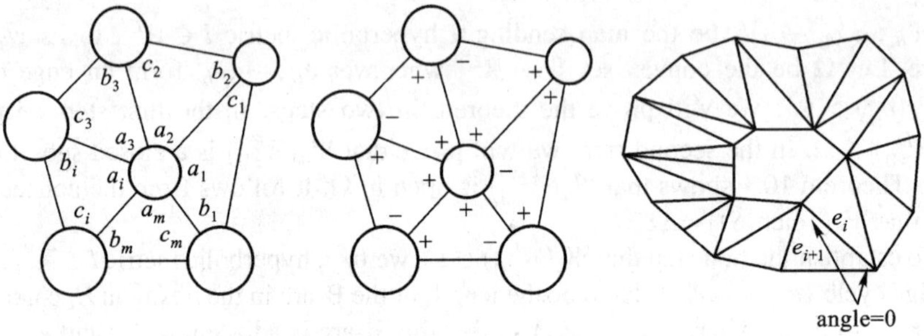

Fig. 10.1. Edge cycles.

We claim that these two rules are contradicting to each other on an ideally triangulated surface (S,T). Indeed, if both (I) and (II) hold, there is a way to assign $+1$ ($=\infty$) and -1 ($=-\infty$) to some corners of a closed triangulated surface so that (i) if a corner in a triangle is assigned -1, then the other two corners are assigned $+1$, and (ii) if c, c' are two corners facing an edge, then the assigned numbers are of opposite signs. Now take a triangle Δ_1 with three corners assigned $+1, +1, -1$. Let v be the vertex of the corner which is assigned -1 and consider all triangles $\Delta_1, \cdots, \Delta_k$, cyclically ordered, having v as a vertex. Using Rules I and II, we conclude that for all $i \geqslant 2$, three corners of Δ_i are assigned $+1, +1, -1$ so that the corner having v as a vertex is assigned $+1$. However, applying this to $\Delta_{k+1} = \Delta_1$, we obtain a contradiction that the corner at v is assigned -1 to start with (see Fig. 10.1).

The above argument shows that if the r-coordinate of a B-arc in the metrics $l^{(m)}$ is not bounded, then we obtain a contradiction.

Case 2. There exists an edge e so that $a(e) = \infty$ and all r-coordinates of B-arcs in the metrics $l^{(m)}$ are uniformly bounded. Let H be a hexagon containing the edge e. By the assumption $\lim_m l^{(m)}(e) = \infty$ and Lemma 10.6(a), (c), there exists an edge e' in H and an B-arc x in H adjacent to e and e' so that

(1) The length of x in $l^{(m)}$ tends to 0 as m tends to ∞ (by Lemma 10.6(a) applied to $l^{(m)}(e) \to \infty$), and

(2) $\lim_m l^{(m)}(e') = \infty$ (by Lemma 10.6(c) applied to $l^{(m)}(x) \to 0$).

By keeping use of these two properties, we obtain a cycle of edges, say $\{e_{n_1}, \cdots, e_{n_k}\}$ so that

(1) $\lim_m l^{(m)}(e_{n_i}) = \infty$,

(2) e_{n_i} and $e_{n_{i+1}}$ lie in a hexagon H_i so that $e_{n_{k+1}} = e_{n_1}$ for $i = 1, 2, \cdots, k$,

(3) the length $a_i^{(m)}$ of the B-arc in H_i adjacent to e_{n_i} and $e_{n_{i+1}}$ converges to 0 (in the metrics $l^{(m)}$).

By definition, the sum of the ψ_h curvature at $e_{n_1}, \cdots, e_{n_k}$ is

$$\sum_{i=1}^{k} z(e_{n_i}) = \sum_{i=1}^{k} \int_0^{\frac{a_i^{(m)}+b_i^{(m)}-c_i^{(m)}}{2}} \cosh^h t dt + \int_0^{\frac{a_i^{(m)}-b_i^{(m)}+c_i^{(m)}}{2}} \cosh^h t dt$$

where $b_i^{(m)}$ and $c_i^{(m)}$ are the lengths of the B-arcs. Due to the assumption that all r-coordinates are bounded in the metrics $l^{(m)}$, the lengths of all B-arcs are bounded. In particular, $b_i^{(m)}$ and $c_i^{(m)}$ are bounded. Thus, as m tends to infinity, due to $\lim_m a^{(m)} = 0$, the limit of the summation $\lim_m \sum_{i=1}^k z(e_{n_i}) = 0$. This contradicts the assumption that $b \in \Omega$ since $\sum_{i=1}^k b(e_{n_i}) > 0$. Thus we conclude that $\Psi_h(\mathbb{R}^E_{>0})$ is closed in Ω.

11

Surface Ricci Flow

In this chapter, we introduce several important concepts in the theory of *surface Ricci flow*.

11.1 Conformal Deformation

Gauss-Bonnet theorem bridges the Gaussian curvature to the topology of a surface. We can consider a special Riemannian metric such that the Gaussian curvature function is constant everywhere, the constant is one of $+1, 0, -1$. The Uniformization theorem 11.11 provides the existence and the uniqueness of such a special metric, as follows.

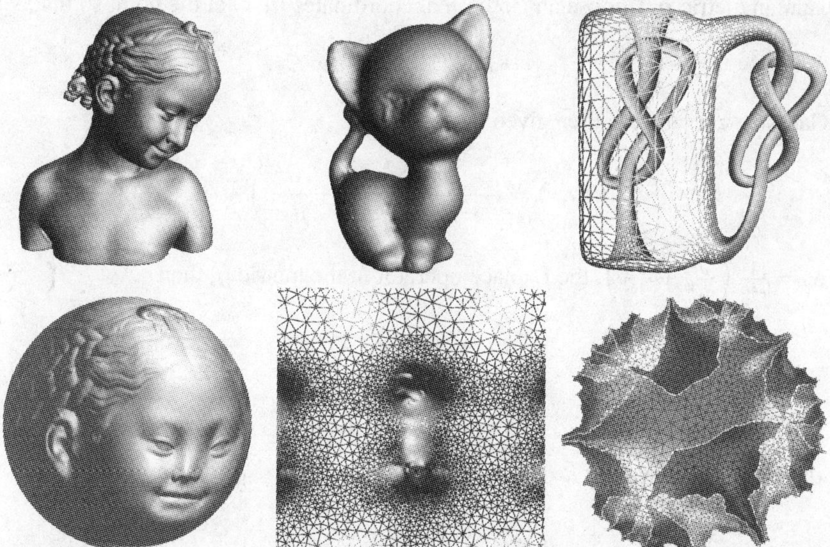

(a) $\chi > 0$ (b) $\chi = 0$ (c) $\chi < 0$

Fig. 11.1. Uniformization theorem: all surfaces with Riemannina metric can be conformally embedded onto three canonical spaces: sphere, plane and hyperbolic space.

Theorem 11.1 (Uniformization theorem). *Let (S,g) be a closed 2-dimensional surface with a Riemannian metric g, then there is a metric $\bar{g}$ conformal to g with constant $+1,-1$ or 0 Gaussian curvatures. Furthermore, the metric $\bar{g}$ is unique if the Euler characteristic of S is negative.*

Such a metric is called the *uniformization metric*. According to Gauss-Bonnet theorem, the sign of the constant Gaussian curvatures is determined by the sign of the Euler characteristic of the surface.

Fig. 11.1 illustrates the Uniformization theorem. The universal covering space of any closed surface with the uniformization metric can be isometrically embedded onto one of the three canonical spaces:

1. the unit *sphere* $\mathbb{S}^2$ for genus zero surfaces with $\chi > 0$;
2. the *plane* $\mathbb{R}^2$ for genus one surfaces with $\chi = 0$;
3. the *hyperbolic space* $\mathbb{H}^2$ for high genus surfaces with $\chi < 0$.

The uniformization metric in Theorem 11.1 is unique up to scaling for Euclidean case, and is unique up to Möbius transformation for the sphere and real projective space.

Definition 11.2 (Conformal deformation). *Let (M,g) be a two-dimensional smooth Riemannian manifold. If $\tilde{g}$ is another Riemannian metric on M, we say that $\tilde{g}$ is* conformal to *g if and only if there exists a positive function $\rho \in C^\infty(M)$, such that*

$$\tilde{g} = \rho g.$$

We also say $\tilde{g}$ is a conformal deformation *of g.*

Here is a way to calculate the Gaussian curvature. Take a point in a surface with a Riemannian metric g. Choose an isothermal coordinates (u,v) at the point so that,

$$g = e^{2\lambda(u,v)}(du^2 + dv^2).$$

The Gaussian curvature is then given by

$$K(u,v) = -\frac{1}{e^{2\lambda(u,v)}}\left(\frac{\partial^2}{\partial u^2} + \frac{\partial^2}{\partial v^2}\right)\lambda.$$

Use $\Delta_g = \frac{1}{e^{2\lambda}}\left(\frac{\partial^2}{\partial u^2} + \frac{\partial^2}{\partial v^2}\right)$, the Laplace operator in the metric g, then

$$K = -\Delta_g \lambda.$$

If

$$\tilde{g} = e^{2\tilde{\lambda}}(du^2 + dv^2),$$

then the Gaussian curvature $\tilde{K}$ of $\tilde{g}$ is given by,

$$\tilde{K} = -\Delta_{\tilde{g}} \tilde{\lambda}.$$

Let $\tau = \tilde{\lambda} - \lambda$. Then we get

$$\begin{aligned}\tilde{K} &= -\Delta_{\tilde{g}}\tilde{\lambda} \\ &= -\frac{1}{e^{2(\lambda+\tau)}}(\Delta\lambda + \Delta\tau) \\ &= -\frac{1}{e^{2\tau}}\frac{1}{e^{2\lambda}}(\Delta\lambda + \Delta\tau) \\ &= \frac{1}{e^{2\tau}}(K - \Delta_g \tau).\end{aligned}$$

The equation is

$$\tilde{K} = \frac{1}{e^{2\tau}}(K - \Delta_g \tau). \tag{11.1}$$

Similarly, if the surface has boundaries, then the geodesic curvature $\tilde{k}$ of $\tilde{g}$ along the boundary can be calculated by

$$\tilde{k} = \frac{1}{e^{\tau}}(k - \partial_{\mathbf{n}} u), \tag{11.2}$$

where $\mathbf{n}$ is the outwards normal on the tangent planes along the boundary. The prescribing Gaussian curvature problem asks given a function $\tilde{K}$ defined on a surface $(M, \mathbf{g})$ with a Riemannian metric $\mathbf{g}$, if there is a metric $\tilde{\mathbf{g}}$ conformal to $\mathbf{g}$ so that the Gaussian curvature of $\tilde{\mathbf{g}}$ is the given function $\tilde{K}$. This problem is very important for engineering applications.

An efficient algorithm to compute the conformal metric $\tilde{\mathbf{g}}$ with prescribed curvature $\tilde{K}$ is the surface Ricci flow method.

11.2 Surface Ricci Flow

Suppose S is a smooth surface with Riemannian metric $\mathbf{g}$. The Ricci flow is the process to deform the metric $\mathbf{g}(t)$ according to its Gaussian curvature $K(t)$, where t is the time parameter:

$$\frac{dg_{ij}(t)}{dt} = -2K(t)g_{ij}(t). \tag{11.3}$$

The Ricci flow was introduced by Hamilton in [Ham] inspired by the heat flows on manifolds. Suppose $T(t)$ is a temperature field on the surface. The heat diffusion equation is

$$\frac{dT(t)}{dt} = -\Delta_g T(t),$$

where Δ is the Laplace-Beltrami operator associated to the surface metric. The temperature field becomes more and more uniform with the increase of t, and it will become a constant eventually.

In physical sense, the curvature evolution induced by the Ricci flow is exactly the same as heat diffusion on the surface. Indeed, Hamilton proved that,

$$\frac{dK(t)}{dt} = -\Delta_{\mathbf{g}(t)} K(t), \tag{11.4}$$

where $\Delta_{\mathbf{g}(t)}$ is the Laplace-Beltrami operator associated to the metric $\mathbf{g}(t)$. We can simplify the Ricci flow equation (11.3). Let $g(t) = e^{2u(t)} g(0)$, then Ricci flow is

$$\frac{du(t)}{dt} = -2K(t). \tag{11.5}$$

It is very easy to see that equation (11.5) preserves conformal class, i.e., if $\mathbf{g}(t)$ is a solution, then $\mathbf{g}(t)$ and $\mathbf{g}(t')$ are conformal. To study the convergence of the Ricci flow (11.5), Hamilton introduced the *normalized Ricci flow* (11.6) so that the surface area is preserved during the flow. here the normalized Ricci flow equation is:

$$\frac{du(t)}{dt} = -2\left(K(t) - 2\pi\frac{\chi(S)}{A(t)}\right), \tag{11.6}$$

where $A(t)$ is the area of the Riemannian metric $g(t)$. It is easy to see that the solutions of (11.5) and (11.6) are related by scaling factor depending only on the area.

The following theorems postulate that the Ricci flow defined in equation (11.3) is convergent and lead to a conformal uniformization metric.

Theorem 11.3 (Hamilton 1988 [Ham]). *For a closed surface of non-positive Euler characteristic, the normalized Ricci flow will converge to a metric such that the Gaussian curvature is constant everywhere.*

Theorem 11.4 (Chow 1991[Cho]). *For a closed surface of positive Euler characteristic, the normalized Ricci flow will converge to a metric such that the Gaussian curvature is constant everywhere.*

The corresponding metric $g(\infty)$ is the *uniformization metric*.

The Ricci flow can be easily modified to compute a metric with a *prescribed curvature* $\tilde{K}$. Indeed, the modified flow is,

$$\frac{du(t)}{dt} = 2(\tilde{K} - K(t)). \tag{11.7}$$

With this modification, for any given curvature $\tilde{K}$, which is admissible with the Gauss-Bonnet theorem, the solution to equation (11.7) exists for all time and converges, as time tends to infinity, to a metric $g(\infty)$ whose curvature is $\tilde{K}$.

12

Geometric Structure

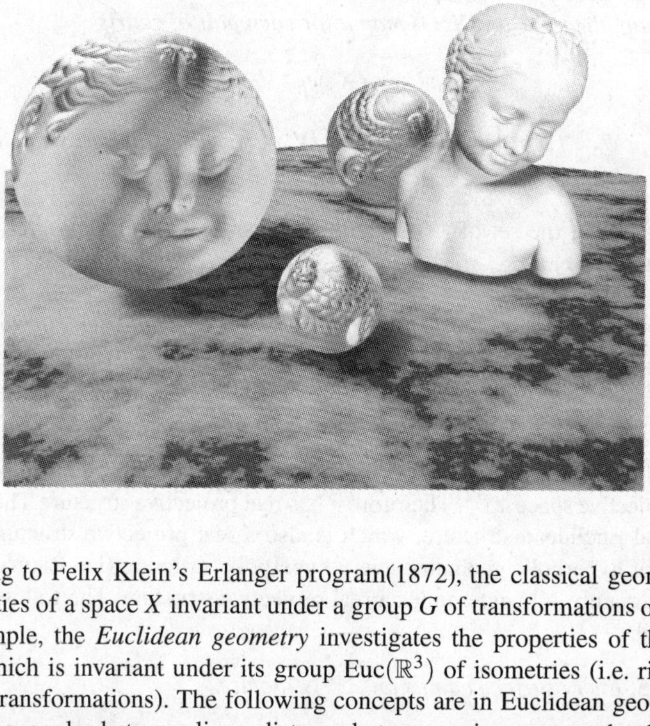

According to Felix Klein's Erlanger program(1872), the classical geometry studies those properties of a space X invariant under a group G of transformations of X.

For example, the *Euclidean geometry* investigates the properties of the Euclidean space $\mathbb{R}^3$, which is invariant under its group $\mathrm{Euc}(\mathbb{R}^3)$ of isometries (i.e. rigid motions, congruence transformations). The following concepts are in Euclidean geometry: parallelism of lines, angles between lines, distance between points, area and volume. All the concepts are derived from the metric structure of the Euclidean space. *Affine geometry* arises when we talk about points, lines and the relation of parallelism. When we only concern points and lines and the relation of incidence between them, then *projective geometry* should be applied.

If M is a manifold, we would like to impart to it different geometries, such as affine geometry and projective geometry. The feasible means are the so-called *affine structure* or *projective structure*.

12.1 (X,G) Geometric Structure

Suppose X is a space with a geometry on it and G is the group of transformations of X, which preserves the geometry.

Definition 12.1 ((X,G) **Atlas**). *Suppose M is a manifold. An (X,G) atlas on M is a collection of atlas $\{(U_\alpha,\phi_\alpha)\}$, such that the local coordinates are defined in X, $\phi_\alpha : U_\alpha \to X$ and $\bigcup_\alpha U_\alpha = M$. If $U_\alpha \cap U_\beta \neq \emptyset$, then the chart transistion function*

$$\phi_\beta \circ \phi_\alpha^{-1} : \phi_\alpha(U_\alpha) \to \phi_\beta(U_\beta)$$

belongs to G.

Definition 12.2 ((X,G)**Structure**). *An (X,G) structure on M is a maximal (X,G)-atlas and (X,G) manifold is a manifold together with an (X,G) structure.*

Definition 12.3 ((X,G) **Map**). *Suppose that M and N are two (X,G) manifolds and $f : M \to N$ is a map, then f is an (X,G) map if for each pair of charts*

$$\phi_\alpha : U_\alpha \to X, \psi_\beta : V_\beta \to X$$

the composition $\psi_\beta \circ f \circ \phi_\alpha^{-1}$ on $\phi_\alpha(U_\alpha \cap f^{-1}(V_\beta))$ is the restriction of an element g in G acting on X.

It is obvious that the set of (X,G) automorphisms of M forms a group, which we denote by $\mathrm{Aut}_{(X,G)}(M)$.

Example 12.4. *The hyperbolic structure is $(\mathbb{H}^2, \mathrm{PSL}(2,\mathbb{R}))$ geometry. The affine structure is $(\mathbb{R}^2, \mathrm{aff}(\mathbb{R}^2))$ geometry where an affine transformation is of the form $x \to Ax + b$ with A in $\mathrm{GL}(2,\mathbb{R})$ and b in $\mathbb{R}^2$. The projective structure is $(\mathbb{RP}^2, \mathrm{PGL}(3,\mathbb{R}))$ geometry where points in $\mathbb{RP}^2$ are lines through the origin of $\mathbb{R}^3$.*

Every surface admits a metric of constant curvature. The sphere $\mathbb{S}^2$ is a covering space of the real projective space $\mathbb{RP}^2$. Therefore it has real projective structure.The plane of $\mathbb{R}^2$ has the natural Euclidean structure, which is also a real projective structure. Using the Klein model of hyperbolic geometry, every hyperbolic structure (i.e. Riemannian metric of constant curvature -1) determines a real projective structure. The following theorem is fundamental.

Theorem 12.5. *Every surface admits an $\mathbb{RP}^2$ structure.*

12.2 Affine Structures on Surfaces

We introduce a basic result of Benzécri that a closed surface admits an affine structure if and only if its Euler characteristic vanishes.

Every orientable noncompact surface admits an immersion into $\mathbb{R}^2$. The immersion determines an affine structure with trivial holonomy. Here is a way to produce the immersion. Suppose M has genus g and with a single boundary. We glue a disk along ∂M to form a closed surface $\bar{M}$. We can construct a canonical homology basis of $\bar{M}$, $\{a_1, b_1, a_2, b_2, \cdots, a_g, b_g\}$. The union of the base curves form a graph, which doesn't intersect ∂M. Then the graph is the deformation retraction of M. We can easily immerse the graph into $\mathbb{R}^2$. Therefore, M can be immersed into $\mathbb{R}^2$ as well. For general non-compact surfaces, we can construct immersion in the same way.

Theorem 12.6 (Benzécri 1955). *Let M be a closed 2-dimensional affine manifold. Then $\chi(M) = 0$.*

Benzécri gave an elementary proof, which can be found in [Gol]. Milnor observed that this result follows from a more general theorem on flat vector bundles [Mil]. Let E be the 2-dimensional oriented vector bundle over M, whose total space is the quotient of $\tilde{M} \times \mathbb{R}^2$ by the diagonal action of π, the fundamental group of the surface, by deck transformations on $\tilde{M}$ and via $L \circ h$ on $\mathbb{R}^2$, where L is the holonomy representation, (i.e. the flat vector bundle over M associated to the linear holonomy representation). Suppose $(p, v) \in \tilde{M} \times \mathbb{R}^2$, then

$$(p,v) \sim (\gamma(p), h(\gamma)(v)), \quad \forall \gamma \in \pi,$$

where γ is an element in the fundamental group π. We also use γ to represent the corresponding deck transformation, $h(\gamma)$ is the corresponding affine transformation. Then the vector bundle is defined as

$$\tilde{M} \times \mathbb{R}^2 / \sim .$$

This bundle has a natural flat structure, since the coordinate changes for this bundle are locally constant linear maps. Now an oriented $\mathbb{R}^2$ bundle over a space M is classified by its Euler class which lies in $H^2(M, \mathbb{Z}) \cong \mathbb{Z}$. If ξ is an oriented $\mathbb{R}^2$ bundle over M which admits a flat structure, Milnor showed that its Euler class

$$|e(\xi)| < g.$$

If M is an affine manifold, then the bundle E is isomorphic to the tangent bundle of M and hence has Euler number $e(TM) = 2 - 2g$. Therefore, the only closed orientable surface whose tangent bundle has a flat structure is a torus. Furthermore, Milnor showed that any $\mathbb{R}^2$ bundle over M whose Euler number satisfies the above inequality has a flat connection.

12.3 Spherical Structure

Suppose M is a closed genus zero Riemann surface. Then M is conformal equivalent to the unit sphere $\mathbb{S}^2$ by the uniformization theorem. Suppose $\phi : M \to \mathbb{S}^2$ is a degree one harmonic map, then ϕ must be conformal. Then ϕ induces a spherical structure on M. Fig. 12.1 illustrates the spherial structure of a sculpture surface.

To compute the map ϕ, let us assume that the 2-sphere is embedded in $\mathbb{R}^3$. Then we can compute the harmonic map in the following way. First we compute the Gauss map of M, $G : M \to \mathbb{S}^2$, send a point $\boldsymbol{r}$ to the normal vector $\boldsymbol{n}$ at $\boldsymbol{r}$, which is a degree one map. Let $f = (f_1, f_2, f_3), f = G$. We treat $f : M \to \mathbb{R}^3$ as the map to $\mathbb{R}^3$.

Then we minimize the harmonic energy of the map,

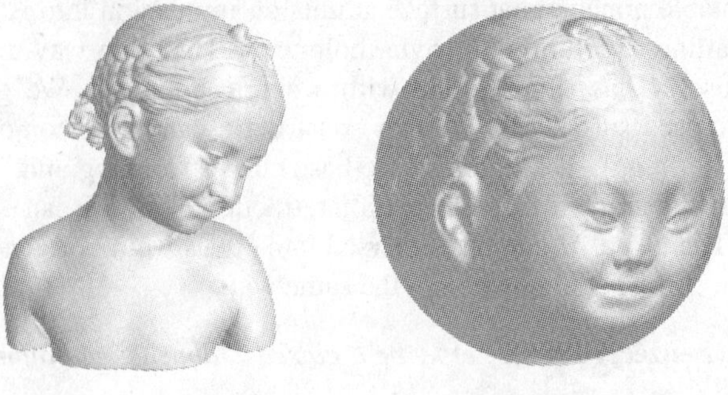

Fig. 12.1. Spherical structure.

$$E(f) = \sum_i \int_M |\nabla f_i|^2 dA.$$

The minimization can be carried out by the following heat flow method

$$\frac{d\boldsymbol{f}}{dt} = -(\Delta \boldsymbol{f} - (\Delta \boldsymbol{f})^{\perp}),$$

where Δ is the Laplace-Beltrami operator on M, $(\Delta \boldsymbol{f})^{\perp}$ is the component of $\Delta \boldsymbol{f}$ along the normal at $\boldsymbol{f}(p), p \in M$. Because the conformal map is not unique, two conformal maps from M to $\mathbb{S}^2$ differ by a Möbius transformation. We need to add some normalization condition. A common method is to fix the images of three points. This variational method leads to the conformal spherical map.

Noncompact orientable surfaces admit spherical structures, because they can be immersed onto the Euclidean plane, which can be mapped onto the sphere by the stereographic projection.

12.4 Euclidean Structure

If a closed orientable surface admits a Euclidean structure, then we can define a flat Riemannian metric on it, which induces zero Gauss curvature everywhere. According to the Gauss-Bonnet theorem, the surface must be a torus. Fig. 12.2 shows the Euclidean structure of a genus one Kitten surface.

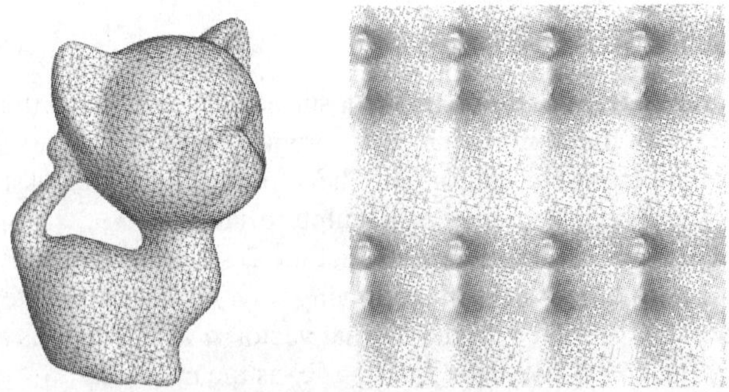

Fig. 12.2. Euclidean structure.

Given a torus, it is straightforward to compute a Euclidean structure on it. There are two direct methods: one is to use a holomorphic 1-form; the other is to use the surface Ricci flow.

Ricci flow method

The second method is to use the surface Ricci flow. Let $\mathbf{g}(t) = e^{2u(t)}\mathbf{g}(0)$, the normalized Ricci flow is

$$\frac{du(t)}{dt} = -2\left(K(t) - 2\pi\frac{\chi(S)}{A(t)}\right)$$

where $A(t)$ is the area under $\mathbf{g}(t)$. In the case of torus, $\chi(S) = 0$. Then the solution is a flat metric on the surface M. Assume $(\tilde{M}, \mathbf{p})$ is the universal covering space of M. Then the projection map $\mathbf{p}$ induces the flat metric to $\tilde{M}$. We can directly isometrically embed $\tilde{M}$ onto the Euclidean plane $\mathbb{R}^2$.

Suppose M is a compact surface with Riemannian metric. Then it is a Riemann surface with induced conformal structure. Let ω be a holomorphic 1-form on M, which induces another flat metric $\tilde{g}$ with cone singularities,

$$\tilde{g} = \omega\bar{\omega}.$$

The zeros of ω become cone singularities. The Gauss curvature is pulse function. Furthermore,

$$\forall p, \quad \omega(p) = 0, \quad K_{\tilde{g}}(p) = 2n\pi, \quad n \in \mathbb{Z}^-.$$

From above discussion, we see that the elements in the holonomy group are rotation free.

Theorem 12.7. *Suppose M is a compact Riemann surface, g is a flat metric with cone singularities on M. Then the curvatures at the singularities are $2n\pi, n \in \mathbb{Z}^-$, if and only if g is induced by a holomorphic 1-form ω.*

Proof. We have shown that a holomorphic 1-form induces a flat metric with cone singularities, which induces a Euclidean structure, so that the holomony group consists of only translations.

Now, assume we have a Euclidean structure which has a rotation free holomony group. Let dz be the holomorphic 1-form on $\mathbb{R}^2$, and also on $\tilde{M}$ by the developing map. Because all deck transformations are translation, dz is well defined on $\tilde{M}/\pi(M)$, which is a holomorphic 1-form on M. □

12.5 Hyperbolic Structure

For closed surfaces with negative Euler numbers, they admit hyperbolic structures as shown in Corollary 9.6. Suppose the surface is equipped with an initial metric $\mathbf{g}_0$. We can compute the uniformization hyperbolic metric using the surface Ricci flow method. In particular, the uniformization metric is conformal to the initial metric $\mathbf{g}_0$ and has -1 Gauss curvature. Let $\mathbf{g}(t) = e^{2u(t)}\mathbf{g}(0)$, the normalized Ricci flow is

$$\frac{du(t)}{dt} = -2\left(K(t) - 2\pi\frac{\chi(S)}{A(t)}\right)$$

where $A(t)$ is the area under $\mathbf{g}(t)$, $\chi(S)$ is the Euler characteristic.Then the solution $\mathbf{g}(\infty)$ is the hyperbolic metric on the surface M. Assume $(\tilde{M}, \mathbf{p})$ is the universal covering space of M. Then the projection map p induces the hyperbolic metric to $\tilde{M}$. We can directly isometrically embed $\tilde{M}$ onto the Poincaré disk Δ.

The deck transformations of $\tilde{M}$ are Möbius transformations. Therefore, the hyperbolic structure induces a *complex projective structure* $\mathbb{CP}^1$ of M. Here by a complex projective structure, we mean $(\mathbb{CP}^1, PSL(2,\mathbb{C}))$ geometry. The complex projective space $\mathbb{CP}^1$ is the Riemann sphere acted by the group of Möbius transformations.

Let Σ be a closed Riemann surface of genus g, and $\phi(z)dz^2$ is a holomorphic quadratic differential on M. We can show that each holomorphic quadratic differential on M corresponds to a holomorphic developing map and induces a complex projective structure and vice versa. Therefore all possible complex projective structures are in one to one correspondence to $(\Sigma, \phi(z)dz^2)$. We know the Teichmüller space for genus g closed surface is $3g-3$ complex dimensional. The space of holomorphic quadratic differentials on M is also $3g-3$ complex dimensional. Therefore, the complex projective structures on a genus g closed surface forms a $6g-6$ complex dimensional space.

If we use the Klein model to represent the hyperbolic space, then all the Möbius transformations are real projective transformations. It shows that a hyperbolic structure induces a *real projective structure* $\mathbb{RP}^2$ on M as shown in Fig. 12.3.

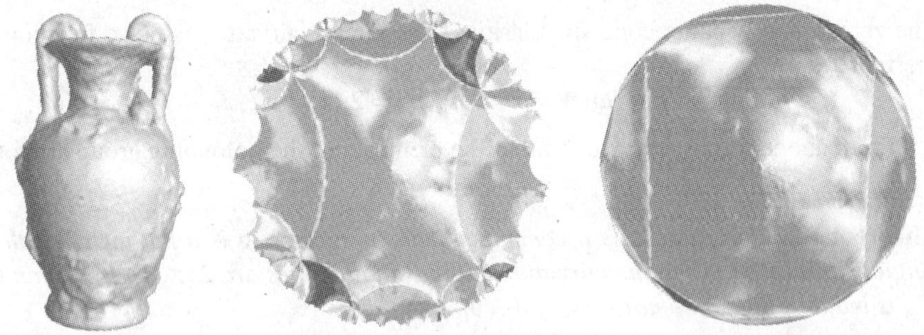

Fig. 12.3. Hyperbolic Structure and induced complex projective structure, real projective structure.

12.6 Real Projective Structure

Let $\mathbb{RP}^2$ be the real projective plane and $PGL(3,\mathbb{R})$ the group of projective transformations $\mathbb{RP}^2 \to \mathbb{RP}^2$. Suppose M is a compact surface. A real projective structure on M is a maximal atlas of coordinate charts locally modeled on $\mathbb{RP}^2$ with coordinate changes lying in $PGL(3,\mathbb{R})$. If M has a $\mathbb{RP}^2$ surface, then a geodesic on M is a curve which in local coordinates maps to a projective line in $\mathbb{RP}^2$.

An $\mathbb{RP}^2$ manifold M can be developed into $\mathbb{RP}^2$ as follows. The coordinate atlas globalizes to define a local diffeomorphism of the universal covering $\tilde{M} \to \mathbb{RP}^2$, called the *developing map*. The deck transformations of the universal covering space $\tilde{M}$ define automorphisms of $\tilde{M}$. The resulting *holonomy homomorphism* $\pi(M) \to PGL(3,\mathbb{R})$ corresponds to the coordinate changes in the atlas for the $\mathbb{RP}^2$ structure. The pair of the developing map and the holonomy homomorphism is unique up to the $PGL(3,\mathbb{R})$ action by composition and conjugation respectively.

The space of projective equivalence classes of real projective structures on a closed orientable surface of genus $g > 1$ is a countable disjoint union of open cells of dimension $16g-16$. This is proved by decomposing a real projective structure into convex subsurfaces along closed geodesics. More references can be found in [Lof] and [Gol].

13

Shape Acquisition and Representation

13.1 Shape Acquisition

Recently, the 3D scanning technology is developing extremely fast. Fig. 13.1 and 13.2 show several facial surfaces with different expressions of the same person, scanned by a scanner based on the phase-shifting method. The scanning speed is as fast as 180 frames per second, each frame has 640×480 samples. The scanner can capture dynamic facial expressions in real time. It is challenging to process the huge amount of geometric data efficiently and robustly. Discrete Ricci flow offers a powerful tool to tackle this problem. The main strategy is to use conformal mappings to transform 3D surfaces to canonical 2D domains, and convert 3D geometric problems to 2D ones.

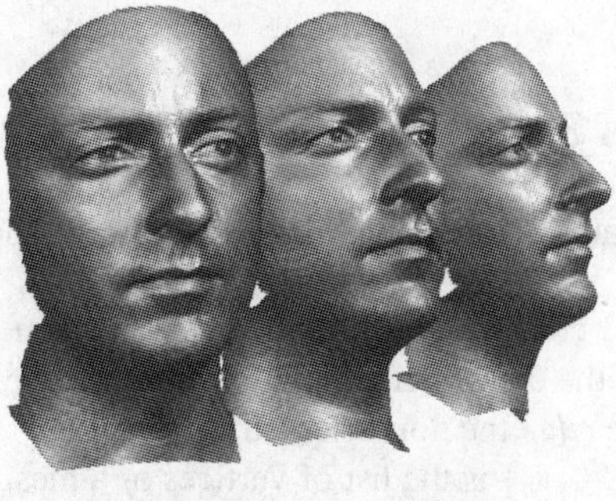

Fig. 13.1. 3D geometry acquired using a real time high speed high resolution scanner.

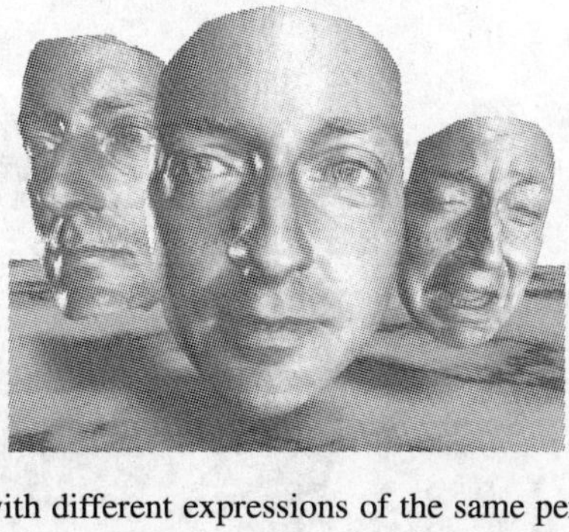

Fig. 13.2. Face geometries with different expressions of the same person, scanned using real time high speed high resolution scanner.

13.2 Triangular Meshes

In digital geometry processing, surfaces are generally represented as triangular meshes, which is the same as triangulation for mathematicians. In computer graphics, triangular meshes are directly supported by graphics hardware. Other geometric representations, such as *splines*, implicit surfaces, are converted to triangular meshes in order to be rendered by graphics hardware.

The popularity of triangular meshes can be explained in terms of theory and practice. In theory, any surface with C^1 smoothness can be triangulated. Simplicial homology and simplicial cohomology theories in topology are directed based on the triangular meshes. In practice, all surfaces in real life can be digitized using $3D$ scanners, and the acquired point clouds can be easily triangulated as meshes. Most software tools support triangular meshes as the default data format. Most importantly, modern graphics hardware supports triangular meshes directly.

A triangular mesh is a 2 dimensional simplical complex in algebraic topology. Basically, a mesh is a set of triangular faces coherently glued together. The topological information is implied by the connectivity. In general, a mesh is embedded in $\mathbb{R}^3$, then each vertex has the Euclidean coordinates in $\mathbb{R}^3$, each face is realized as a planar triangle. Each edge has the length induced by the Euclidean metric of $\mathbb{R}^3$. The mesh is flat almost everywhere, except at the vertices, where the curvatures can be defined and measured. Topological problems can be solved exactly on the mesh, without any approximation. Differential forms, curvatures, geodesics and conformal mappings can be approximated on meshes.

13.3 Half-Edge Data Structure

In order to represent a triangular mesh in geometric software, the *half-Edge* data structure is commonly used.

A mesh $M=(V,E,F)$ consists of a list of *vertices* V, *non-oriented edges (edges)* E and *oriented faces* F. For the convenience of programming, we use the *half-edge* to refer the oriented edges and the *edge* the non-oriented edges.

Suppose $V=\{v_0,v_1,\cdots,v_n\}$ is the list of vertices of a mesh M. Each half-edge $h_{ij}=[v_i,v_j]$ connecting two vertices v_i and v_j, where the *source vertex* is v_i and the *target vertex* is v_j. The boundary of h_{ij} is

$$\partial[v_i,v_j]=v_j-v_i.$$

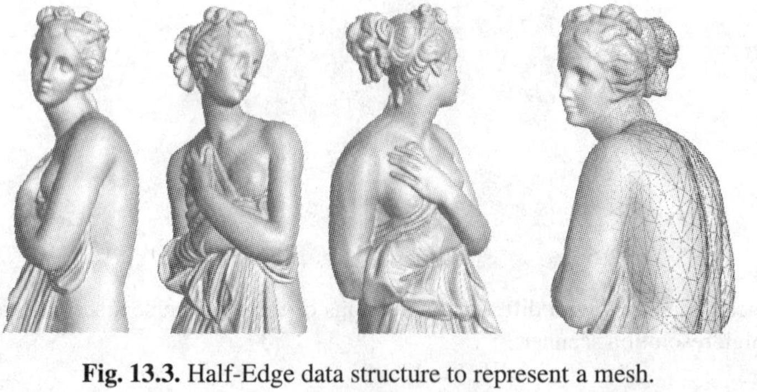

Fig. 13.3. Half-Edge data structure to represent a mesh.

Each oriented face $f_{ijk} = [v_i, v_j, v_k]$ consists of three vertices, v_i, v_j and v_k, ordered counter-clock-wisely with respect to the orientation. Furthermore, the boundary of each face has three consecutive half-edges,

$$\partial[v_i, v_j, v_k] = [v_i, v_j] + [v_j, v_k] + [v_k, v_i].$$

In practice, we mainly consider the *oriented mesh*,

Definition 13.1 (Oriented mesh). *A mesh is orientable, if one can select an orientation for each face, such that each interior edge attaches to 2 half-edges with opposite orientations.*

Intuitively, if the mesh is closed embedded in $\mathbb{R}^3$, the mesh divides the whole space of $\mathbb{R}^3$ into two volumes. One is finite, called the *inside volume*, the other is infinite, called the *outside volume*. The *normal* of an oriented face $f = [v_i, v_j, v_k]$ is the unit vector perpendicular to the face plane and is expressed as

$$\boldsymbol{n} = \frac{(v_j - v_i) \times (v_k - v_i)}{|(v_j - v_i) \times (v_k - v_i)|}. \tag{13.1}$$

Namely, the normal and the $v_j - v_i$, $v_k - v_i$ satisfy right hand rule. If the mesh is orientable, then all the normals are pointing towards outside.

If the mesh is not closed, there will be boundaries. For all interior non-oriented edges e_{ij} there should be two half-edges attaching to it with opposite orientations, $h_{ij} = [v_i, v_j]$ and $h_{ji} = [v_j, v_i]$. All the boundary edges have only one half-edge attaching to them. If we walk along one of the boundaries, which is a list of consecutive half-edges, our walking direction is consistent with the half-edge orientation. The surface is on our left hand side, the hole is on our right hand side.

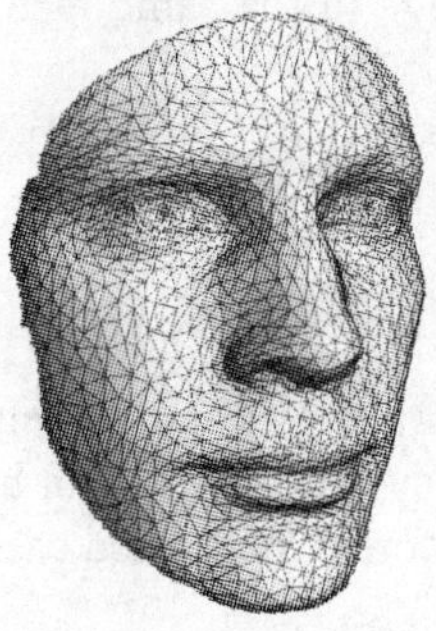

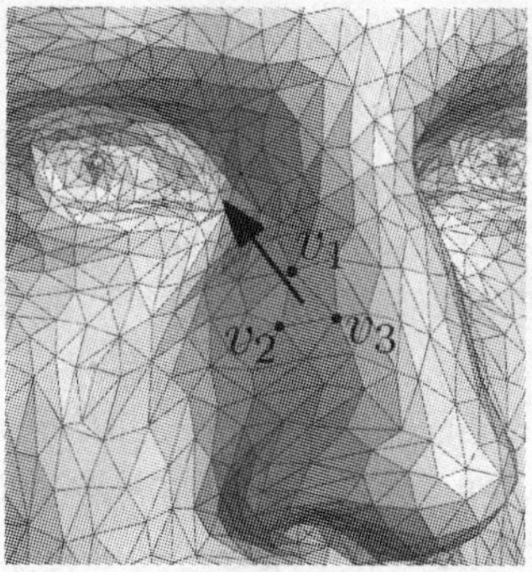

Fig. 13.4. Mesh orientation and surface normal. Three vertices are ordered counter-clock-wisely, the normal is towards the front side. Each interior edge has two half-edges attached with opposite orientation. Each boundary edge has one half-edge attached. If we walk along the boundary, the surface is on our left hand side, the hole is on our right hand side.

Definition 13.2 (Dual half-edge). *Suppose an oriented edge is $h_{ij} = [v_i, v_j]$, then we call half-edge $h_{ji} = [v_j, v_i]$ the* dual half-edge *of h_{ij}.*

In most cases, the surfaces we are processing are orientable. The half-edge data structure representation of an orientable surface has the following property: *The half-edges attach to one edge must be dual to each other.*

All interior edges are adjacent to two faces. They point to two half-edges dual to each other. On the boundary of the mesh, the edge has only one half-edge.

Definition 13.3 (Boundary edge). *Suppose an edge in the mesh points to only one half-edge, then the edge is called a* boundary edge. *The half-edge has no dual half-edge, and is called a* boundary half-edge.

For the convenience of geometric processing, all the adjacent simplexes are connected together coherently using pointers in the data structure. In other words, the topological adjacency information is encoded by the combinatorial structure of the pointers. In practice, it is important to preserve the consistency among the pointers and it is very efficient to execute the following operations using the pointer structure. There are,

- Traverse all the vertices connecting to a given vertex in counter-clock-wise order.
- Traverse all the edges connecting to a given vertex in counter-clock-wise order.
- Traverse all the faces connecting to a given vertex in counter-clock-wise order.
- Traverse all the half-edges pointing towards to a given vertex in counter-clock-wise order.
- Traverse all the half-edges pointing away from a given vertex in counter-clock-wise order.
- Traverse all the half edges of a boundary of the mesh in clock-wise order.

All the classes of half-edge data structure are connected to the half-edge class. Therefore, the half-edge class plays a vital role in the half-edge data structure. The following discussion goes through all the pointers in half-edge data structure:

1. Each half-edge points to the face it belongs to, $h \to f$.
2. Each face points to its first half-edge, $f \to h$.
3. Each half-edge points to the next half-edge in the same face ordered counter-clockwisely, $h \to h$.
4. Each half-edge points to the previous half-edge in the same face ordered counter-clock-wisely, $h \to h$.
5. Each half-edge points to the edge it attaches to, $h \to e$.
6. Each edge points to all half-edges attaching to it, $e \to h$.
7. Each half-edge points to its target vertex, $h \to v$.
8. Each vertex points to the first half-edge which targets at the vertex, $v \to h$.

There are no other direct connections. For example, no pointer connects two vertices, or a vertex and a face. But all the indirect adjacency information can be obtained by the pointers. Here, we discuss two simple examples. More complicated cases can be handled in the same way.

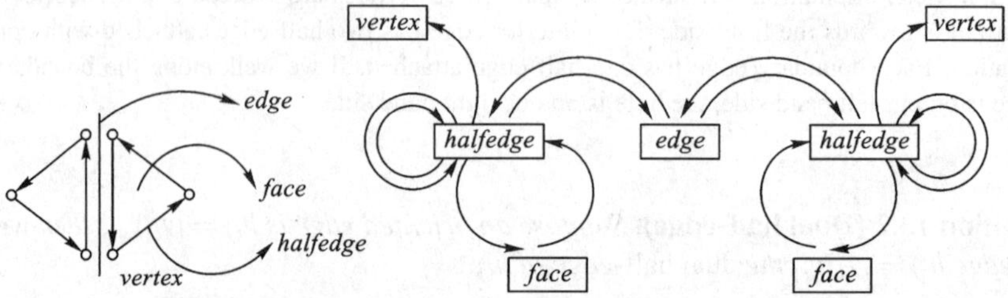

Fig. 13.5. Half edge data structure.

14

Discrete Ricci Flow

This chapter introduces the practical algorithms for simulating smooth surface Ricci flow introduced in Chapter 11. The goal is to construct a Riemannian metric with a prescribed Gaussian curvature.

14.1 Circle Packing Metric

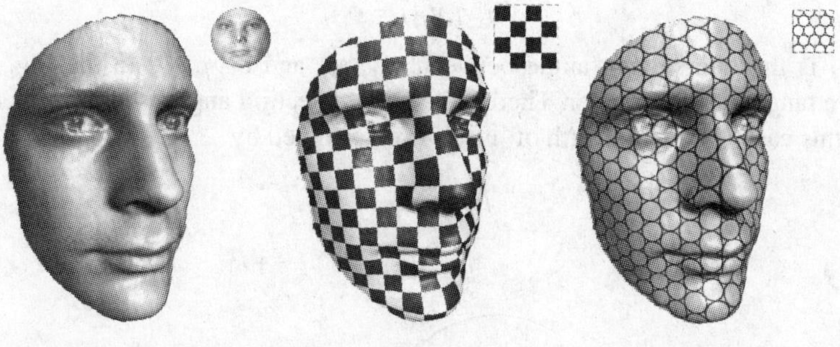

Fig. 14.1. Conformal mappings transform infinitesimal circles to infinitesimal circles, preserve the intersection angles among the circles.

Conformal mappings have a special property that they transform infinitesimal circles to infinitesimal circles, keeping the intersection angles unchanged. Fig. 14.1 visualizes this property. A human face surface is conformally mapped to the planar unit disk by a Riemann mapping, as shown in the first column. We texture map a checkerboard onto the surface as shown in the third column. We can see all the corner angles are well-preserved. If we replace the checkerboard by a *circle packing pattern* as shown in the second column, we can see all circles on the texture are mapped to circles on the surface and all tangency relations among circles are preserved on the surface.

Fig. 14.2 explains the concept of circle packing of tangential type. Suppose T is a planar triangulation, each vertex v_i is associated with a circle $c(v_i, \gamma_i)$, which represents a circle centered at v_i with radius γ_i. Suppose $[v_i, v_j]$ is an edge of T, then two circles $c(v_i, \gamma_i)$ and $c(v_j, \gamma_j)$ are tangent to each other. The more general form of circle packing involves circles intersecting in angles, firstly studied in details by Thurston. Here is the setup. Suppose the vertex set of T is

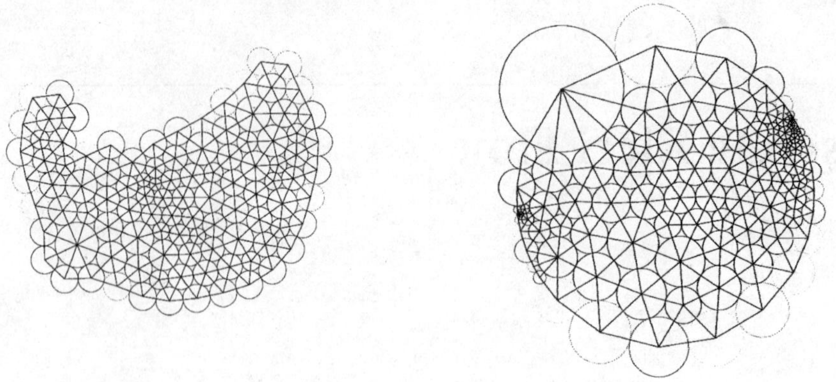

Fig. 14.2. Discrete conformal mapping approximated by circle packing.

$$V = \{v_1, v_2, \cdots, v_n\},$$

and

$$\Gamma : V \to \mathbb{R}^+, \Gamma(v_i) = \gamma_i,$$

represents the circle radii of $c(v_i, \gamma_i)$'s. We assume that the circles intersect in a prescribed angle

$$\Phi([v_i, v_j]) = \phi_{ij},$$

where ϕ_{ij} is the intersection angle between $c(v_i, \gamma_i)$ and $c(v_j, \gamma_j)$. In the Fig. 14.2, all circles are tangent to each other. Therefore, the intersection angles Φ are zeros for every edge. In this case, the edge length of $[v_i, v_j]$ is determined by

$$l_{ij} = \gamma_i + \gamma_j.$$

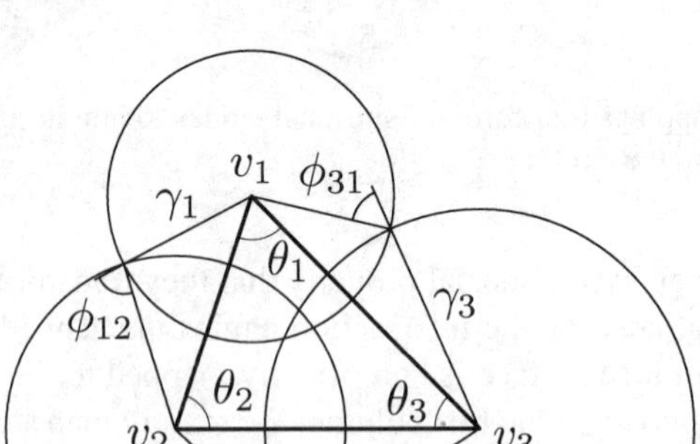

Fig. 14.3. Circle packing metric.

As shown in Fig. 14.3, planar circle packing can be generalized to surface circle packing in the following way:

1. Take a triangulation of a surface of arbitrary topology;
2. The intersection angle Φ is not zero any more, but arbitrary acute angle;
3. Each circle is generalized to a cone centered at the vertex;
4. The edge length is determined by the following Euclidean cosine law

$$l_{ij} = \sqrt{\gamma_i^2 + \gamma_j^2 + 2\cos\phi_{ij}\gamma_i\gamma_j}.$$

In the following, we formally define circle packing metric for a triangular mesh.

Definition 14.1 (Discrete metric). *Suppose M is a triangular mesh, a* discrete metric *is a function defined on non-oriented edges of M,*

$$l : E \to \mathbb{R}^+,$$

such that for each triangle $[v_i, v_j, v_k]$, the triangle inequality holds

$$l([v_i, v_j]) + l([v_j, v_k]) > l([v_k, v_i]).$$

This is the same as Euclidean polyhedral surface on M defined in Definition 4.1. If the mesh is assigned with a circle packing, then it has a natural circle packing metric.

Definition 14.2 (Circle packing metric). *Suppose M is a triangular mesh, with a circle packing (M, Γ, Φ). Then the* circle packing metric *is defined as*

$$l([v_i, v_j]) = \sqrt{\gamma_i^2 + \gamma_j^2 + 2\cos\phi_{ij}\gamma_i\gamma_j}.$$

If all the intersection angles are acute, then Thurston proved that the edge lengths induced by a circle packing satisfy triangle inequality. Fig. 14.4 shows a circle packing on the Stanford bunny surface.

If S is a smooth surface with a Riemannian metric $\mathbf{g} = (g_{ij})$, then a conformal metric is $e^{2u}\mathbf{g}$, where $u : S \to \mathbb{R}$ is a function on the surface. In the discrete case, we can define conformal circle packing metric analogically.

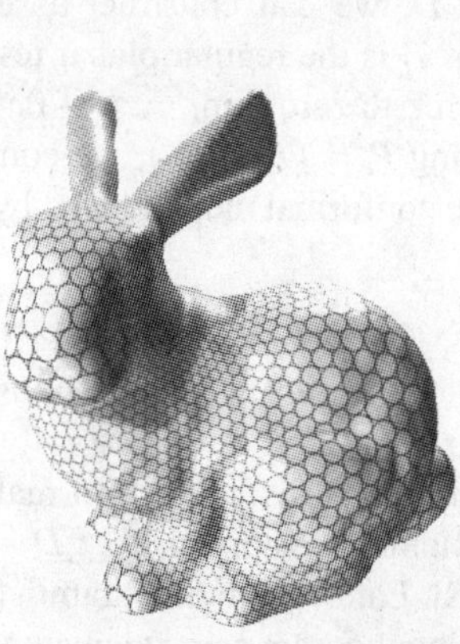

Fig. 14.4. Circle packing on the Stanford Bunny surface.

Definition 14.3 (Conformal circle packing metric). *Suppose M is a triangular mesh with two different circle packing metrics (M,Γ_1,Φ_1) and (M,Γ_2,Φ_2). If*

$$\Phi_1 \equiv \Phi_2,$$

then two circle packing metrics are called conformally equivalent.

Namely, two circle packings with the same intersection angles but different circle radii induce the conformal circle packing metric.

Fig. 14.2 demonstrates two conformal circle packing metrics of the same combinatorial triangulation T. We can define a quasi-conformal map between two circle packing metrics,

$$\phi : (M,\Gamma_1,\Phi_1) \to (M,\Gamma_2,\Phi_1)$$

as follows. Each face $f = [v_0,v_1,v_2]$ has two different embeddings induced by the two metrics. In the first emebedding, We assume the local coordinates of v_i is (x_i,y_i), in the second embedding, the coordinates of v_i is (u_i,v_i), then the map

$$\phi : (x_i,y_i) \to (u_i,v_i)$$

is piecewise linear. The restriction of ϕ on each face is non-degenerated. Hence, a quasi-conformal map is uniquely determined by (Γ_1,Φ_1) and (Γ_2,Φ_1). Carefully examing the figures in Fig. 14.2, we can see that ϕ maps an irregular planar domain $|M|$ to the disk. ϕ is a discrete approximation of the Riemann mapping from the domain $|M|$ to the unit disk.

The first fundamental result for circle packing is the following theorem, proved by Köbe in 1920's and reproved independently by Andreev and Thurston in 1970's.

Theorem 14.4. *Let K be a combinatorial closed disk (simply connected,finite and with nonempty boundary). Then there exists an univalent tangential circle packing $P' \subset \mathbb{D}$ for K, where $\mathbb{D}$ is the planar unit disk, such that every boundary circle is tangent to the unit circle. Two such circle pakcings differ by a Möbius transformation.*

Such a kind of circle packing is called the maximal circle packing of K.

Given a planar domain D, we can construct a sequence of circle packings $P_i = (T_i,\Gamma_i,\Phi_i)$, $\Phi_i \equiv 0$, such that T_i is the regular planar tessellation by equilateral triangles, Γ_i are constant, the radii shrink to zero, $\lim_{k\to\infty} \gamma \to 0$. T_i exhausts D from interior eventually. For each circle packing $P_i = (T_i,\Gamma_i,\phi_i)$, we compute its maximal circle packing $\tilde{P}_i = (T_i,\tilde{\Gamma}_i,\Phi_i)$. The discrete conformal map induced by the two circle packings is

$$\psi_i : P_i \to \tilde{P}_i.$$

Furthermore, suppose $p,q \in D$. We use a Möbius transformation to ensure $\psi_i(p)$ is the origin, $\psi_i(q)$ is on the real axis.

Thurston [Thu] conjectured that discrete conformal map ψ_i induced by circle packing converges to the real Riemann mapping $\psi_\infty : D \to \mathbb{D}$. Rodin and Sullivan proved Thurston's conjecture in [RS]. Later, He and Schramm [HS] used circle packing to prove the classical Riemann mapping theorem using elementary geometry. The whole literature of conventional complex analysis can be based on circle packing [St].

In the following discussion, we generalize the circle packing method from domains on complex plane to general Riemann surfaces shown in Fig. 14.5.

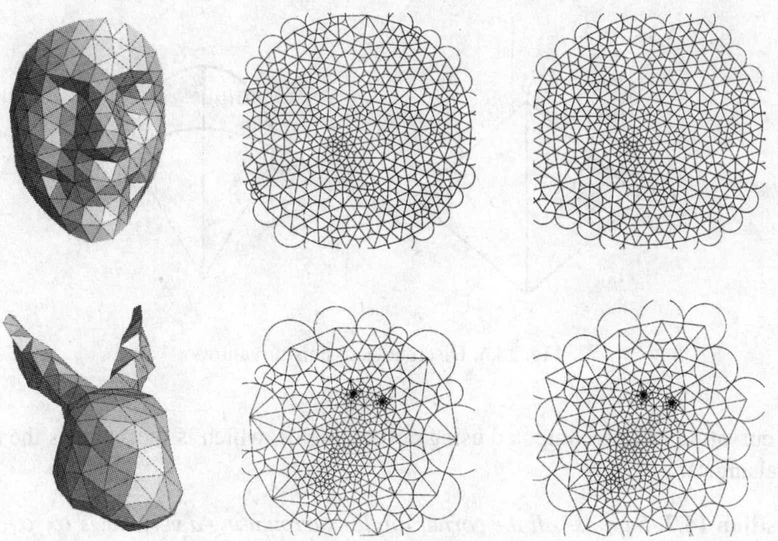

Fig. 14.5. Approximating conformal mappings by circle packing. The circle radii are changed while tangency relations are preserved. The second column shows a circular boundary condition; the third column shows a free boundary condition. For the free boundary condition, all circle radii of boundary vertices are equal.

14.2 Discrete Gaussian Curvature

In the smooth case, Gaussian curvature can be defined by means of the Gauss map. Suppose $G: S \to \mathbb{S}^2, p \to \boldsymbol{n}(p)$ is the Gauss map of the surface, which maps a point $p \in S$ to its normal $\boldsymbol{n}(p)$. A surface patch D is chosen, which is a neighborhood of the point p, $p \in D$. $G(D)$ is a region on the unit sphere $\mathbb{S}^2$. Then the Gauss curvature at p is the limit of the area ratio between $G(D)$ and D, when D shrinks to p.

The discrete Gauss map of a mesh is defined using the face normal.

Definition 14.5 (Discrete Gauss map). *Suppose M is a triangular mesh, embedded in $\mathbb{R}^3$. Then each face f has a unique normal $\boldsymbol{n}(f)$. For each point $p \in f$, and $p \notin \partial f$, the Gauss map is defined as*

$$G(p) = \boldsymbol{n}(f).$$

For points on edges, the Gauss map is not defined.

Fig. 14.6 illustrates the definition of discrete Gauss curvature.

Definition 14.6 (Discrete Gaussian curvature). *Suppose v is an interior vertex of M, $v_0, v_1, \cdots, v_n$ are all neighboring vertices of v, sorted counter-clock-wisely. The normal of face $[v, v_k, v_{k+1}]$ is denoted as $\boldsymbol{n}_k$. Then we consider the spherical polygon*

$$G(v) = [\boldsymbol{n}_0, \boldsymbol{n}_1, \cdots, \boldsymbol{n}_n].$$

All the edges are great arcs of the sphere. The order of the $\boldsymbol{n}_i$'s might be counter clock wisely or clock wisely, determined by the local geometry of v, then we define the discrete Gauss curvature of v as the area of the spherical polygon, denoted as $K(v)$.

If the vertices of $G(v)$ are sorted counter clock wisely, then the oriented area is positive. Otherwise, the oriented area is negative. In the following, we show that the discrete

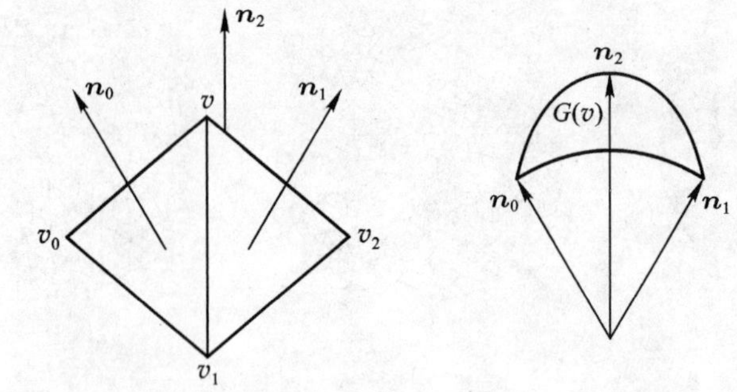

Fig. 14.6. Discrete Gaussian curvature

Gauss curvature can be computed using corner angles, which is the same as the exterior dihedral angles.

Proposition 14.7. *Suppose all the corner angles surrounding a vertex v is $\alpha_0, \alpha_1, \cdots, \alpha_n$, then*

$$K(v) = 2\pi - \sum_{k=0}^{n} \alpha_k.$$

Proof. As shown in Fig. 14.6, the normals $\boldsymbol{n}_k$ and $\boldsymbol{n}_{k+1}$ and the great arc connecting them $\overline{\boldsymbol{n}_k \boldsymbol{n}_{k+1}}$ on the unit sphere form a plane π_k. The edge $[v, v_{k+1}]$ is orthogonal to both $\boldsymbol{n}_k$ and $\boldsymbol{n}_{k+1}$. Therefore, $[v, v_{k+1}]$ is orthogonal to the plane π_k. And $[v, v_{k+1}]$ is orthogonal to the tangent vector at $\boldsymbol{n}_{k+1}$ to the arc $\overline{\boldsymbol{n}_k \boldsymbol{n}_{k+1}}$. Similarly, edge $[v, v_{k+2}]$ is orthogonal to the tangent vector at $\boldsymbol{n}_{k+1}$ to the arc $\overline{\boldsymbol{n}_{k+1} \boldsymbol{n}_{k+2}}$.

Therefore the edges of the angle $\alpha_{k+1} = \angle v_{k+1} v v_{k+2}$ are perpendicular to the edges of the corner angle at $\boldsymbol{n}_{k+1}$. And α_{k+1} equals π minus the $(k+1)$-th inner angle of $G(v)$. Hence α_{k+1} equals the $(k+1)$-th exterior angle of $G(v)$.

According to Gauss-Bonnet theorem

$$\int_{G(v)} dA + \sum \alpha_i = 2\pi,$$

we have $K(v) = \int_{G(v)} dA = 2\pi - \sum \alpha_i$. Compare Corollary 2.5 in which we use dihedral angles. □

The Gauss-Bonnet theorem bridges differential geometry and topology. It also holds for discrete mesh.

Theorem 14.8 (Discrete Gauss-Bonnet theorem). *Suppose M is a closed triangular mesh with a discrete metric. Then*

$$\sum_{v \in M} K(v) = 2\pi \chi(M),$$

where $\chi(M)$ is the Euler number of the mesh, $\chi(M) = V + F - E$.

Proof. The summation of three angles of each face equals π. The total corner angles of a mesh equals

$$\sum \alpha = F\pi.$$

The total discrete Gauss curvature is

$$\sum_{v \in M} K(v) = \sum_{v \in M} 2\pi - \sum \alpha = 2V\pi - F\pi.$$

Each face has three edges, and each edge is shared by two faces,

$$3F = 2E.$$

Therefore

$$\sum_{v \in M} K(v) = 2\pi(V + F - E). \qquad \square$$

Now, we generalize the above results to meshes with boundaries. First, we define Gauss curvature of boundary vertices.

Definition 14.9 (Discrete Gaussian curvature of boundary vertex). *Suppose M is a mesh with boundaries and $v \in \partial M$. If the summation of all the corner angles surrounding v is α, then the discrete Gauss curvature at v is*

$$K(v) = \pi - \sum \alpha.$$

The discrete Gauss-Bonnet theorem still holds for meshes with boundaries. Let $\bar{M}$ is the double covering of M. Then the following relations hold

$$\sum_{v \in M} K(v) = \frac{1}{2} \sum_{\bar{v} \in \bar{M}} K(\bar{v}),$$

also

$$\chi(M) = \frac{1}{2}\chi(\bar{M}).$$

This result shows that the total Gauss curvature is independent of the choice of the metric.

14.3 Discrete Surface Ricci Flow

Suppose S is a smooth surface with a Riemannian metric $\boldsymbol{g} = (g_{ij})$. The normalized Ricci flow deforms the metric $\boldsymbol{g}(t)$ according to the Gaussian curvature $K(t)$ of $\boldsymbol{g}(t)$, where t is the time parameter. Let $\boldsymbol{g}(t) = e^{2u(t)}\boldsymbol{g}(0)$, then the normalized Ricci flow is

$$\frac{du(t)}{dt} = -2(K(t) - 2\pi\frac{\chi(S)}{A(t)}). \tag{14.1}$$

where $A(t)$ is the area of the Riemannian metric $\boldsymbol{g}(t)$. Normalized Ricci flow preserves the area during the flow.

In a physical sense, the curvature evolution induced by the Ricci flow is exactly the same as the heat diffusion on the surface. To be precise,

$$\frac{dK(t)}{dt} = -\Delta_{g(t)}K(t),$$

where $\Delta_{g(t)}$ is the Laplace-Beltrami operator induced by the metric $\boldsymbol{g}(t)$.

The Ricci flow can be easily modified to compute a metric with a prescribed curvature $\bar{K}$. The modified flow is

$$\frac{du(t)}{dt} = 2(\bar{K} - K(t)). \tag{14.2}$$

The discrete surface Ricci flow is the exact analogy of the smooth surface Ricci flow as described in equation (14.2). It was introduced by Chow and Luo in 2003, [CL]. Suppose we are given a mesh M with a circle packing metric (M, Γ, Φ). The circle centered at the vertex v_i has radius γ_i.

If the current Gaussian curvature of v_i is K_i and the prescribed Gaussian curvature is $\bar{K}_i$, then the discrete Ricci flow is defined as

$$\frac{du_i(t)}{dt} = (\bar{K}_i - K_i(t)), \tag{14.3}$$

where $u_1 = \ln \gamma_i$.

Definition 14.10 (Convergence). *A solution to equation (14.3)* exists *and is* convergent, *if*

1. $\lim_{t\to\infty} K_i(t) = \bar{K}_i, \forall i,$
2. $\lim_{t\to\infty} \gamma_i(t) = \bar{\gamma}_i \in \mathbb{R}^+, \forall i.$

A solution converges exponentially *if there are constants c_1, c_2, so that for all time $t \geqslant 0$,*

$$|K_i(t) - \bar{K}_i| \leqslant c_1 e^{-c_2 t},$$

and

$$|\gamma_i(t) - \bar{\gamma}_i| \leqslant c_1 e^{-c_2 t}.$$

Chow and Luo proved that the discrete Ricci flow is exponentially convergent in [CL].

Theorem 14.11. *Suppose (M, Φ) is a closed weighted mesh, whose $\Phi \in [0, \frac{\pi}{2}]$. Given any initial circle packing metric based on the weighted mesh, the solution to the discrete Ricci flow (14.3) in the Euclidean geometry with the given initial value exists for all time and converges exponentially fast.*

For a given weighted mesh (M, Φ), we say a Gaussian curvature $K : V \to \mathbb{R}$ is *admissible*, if there exists a circle packing metric (M, Φ, Γ), whose Gaussian curvature equals K.

Theorem 14.12. *Suppose (M, Φ) is a weighted mesh, $\Phi : E \to [0, \frac{\pi}{2}]$, I is a subset of vertices V. Then*

$$\sum_{i\in I} K_i > - \sum_{(e,v)\in Lk(I)} (\pi - \Phi(e)) + 2\pi\chi(F_I).$$

where F_I is the subcomplex consisting of 2-cells whose vertices are in I and $Lk(I) = \{(e,v) | e \text{ is an edge so that } e \cap I = \emptyset \text{ and the vertex } v \in I \text{ so that } e, v \text{ form a triangle}\}$.

The Algorithm 1 below describes the process to compute discrete surface Ricci flow in details. In practice, if the desired metric is required to be conformal to the initial induced Euclidean metric, the initial circle packing metric need to be carefully chosen. If the quality of the triangulation of the mesh is not good, such as too many skinny triangles or obtuse angles, it is difficult to find a good initial circle packing metric. Therefore, sometimes re-meshing is necessary to improve the quality of triangulations. We find the initial circle packing metric that is close to the induced Euclidean metric. At each step, we add one normalization step to ensure

$$\sum_{v_i \in M} \log \gamma_i = 0.$$

This is equivalent to the area preserving constraint in the smooth surface Ricci flow.

Algorithm 1: Discrete surface Ricci flow.

input : A mesh M embedded in $\mathbb{R}^3$, target curvature $\bar{K}$, curvature error threshold ε
output: a circle packing metric (M,Γ,Φ) which induces $\bar{K}$.

Compute the initial circle packing metric (M,Γ_0,ϕ);
while $max|K_i - \bar{K}_i| > \varepsilon$ **do**
  forall *edge* $e = [v_i, v_j] \in M$ **do**
    Compute the edge length

$$l_{ij} \leftarrow \sqrt{\gamma_i^2 + \gamma_j^2 + 2\gamma_i\gamma_j \cos\phi_{ij}}$$

  end
  forall *Corner in mesh* $c_i \in [v_i, v_j, v_k]$ **do**
    Compute the corner angle c_i

$$c_i \leftarrow \cos^{-1} \frac{l_{ij}^2 + l_{ki}^2 - l_{jk}^2}{2l_{ij}l_{ik}}.$$

  end
  forall *Vertex* $v_i \in M$ **do**
    Compute the Gaussian curvature of v_i;

$$K_i \leftarrow 2\pi.$$

    forall *Corners c attached to* v_i **do**

$$K_i \leftarrow K_i - c.$$

    end
  end
  forall *Vertex* $v_i \in M$ **do**

$$\gamma_i \leftarrow \gamma_i + \gamma_i \times (\bar{K}_i - K_i)$$

  end

$$s \leftarrow 0,$$

  forall *Vertex* $v_i \in M$ **do**

$$s \leftarrow s + \log \gamma_i,$$

  end
  forall *Vertex* $v_i \in M$ **do**

$$\gamma_i \leftarrow \gamma_i \times \exp -\frac{s}{|V|}.$$

  end
end

14.4 Newton's Method

The discrete surface Ricci flow in equation (14.3) is in fact the negative gradient flow of some convex energy. The desired circle packing metric is the optimum point of the energy. In this section, we explicitly formulate the energy and its Hessian matrix. The optimum can be computed by using Newton's method with much higher efficiency.

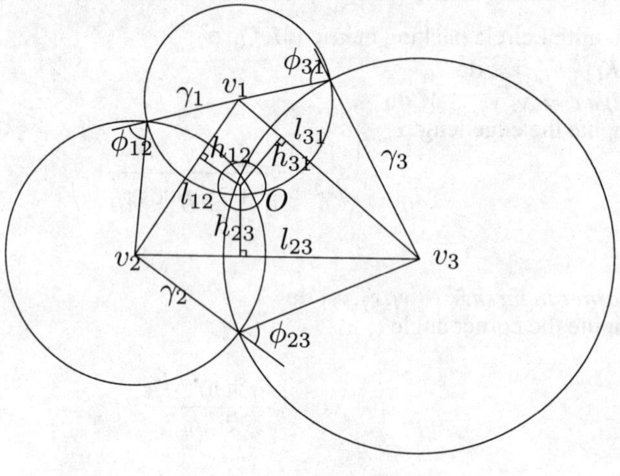

Fig. 14.7. Circle packing metric on a triangle.

Fig. 14.7 illustrates the circle packing metric of a triangle $[v_1, v_2, v_3]$. Three circles $c(v_i, \gamma_i)$ are centered at each vertex, each pair has a common chord. Three common chords intersect at point O, called the *center* of the triangle. The center can be obtained in another way. There exists a unique circle orthogonal to the three circles, the center of the circle is also at O. The distance from O to the edge $[v_i, v_j]$ is h_{ij}.

Let $u_i = \log \gamma_i$. By a direct calculation, it can be verified that

$$\frac{\partial \theta_j}{\partial u_i} = \frac{\partial \theta_i}{\partial u_j} = h_{ij}.$$

Because of $\theta_1 + \theta_2 + \theta_3 = \pi$,

$$\frac{\partial \theta_i}{\partial u_i} = -\frac{\partial \theta_j}{\partial u_i} - \frac{\partial \theta_k}{\partial u_i} = -\frac{\partial \theta_i}{\partial u_j} - \frac{\partial \theta_i}{\partial u_k}. \tag{14.4}$$

Now we define the edge weight w_{ij} in the following way: suppose two faces $[v_i, v_j, v_k]$ and $[v_j, v_i, v_l]$ share an edge $[v_i, v_j]$, O_k is the center of $[v_i, v_j, v_k]$ and O_l is the center of $[v_j, v_k, v_l]$, h_{ij}^k is the distance from O_k to the edge $[v_i, v_j]$, h_{ji}^l is the distance from O_l to the edge $[v_j, v_i]$. Then

$$w_{ij} = \frac{h_{ij}^k}{l_{ij}} + \frac{h_{ji}^l}{l_{ij}}.$$

If $[v_i, v_j]$ is a boundary edge and only attaches to one face $[v_i, v_j, v_k]$, then

$$w_{ij} = \frac{h_{ij}^k}{l_{ij}}.$$

From equation (14.4), it is easy to show that

$$\frac{\partial K_i}{\partial u_j} = \frac{\partial K_j}{\partial u_i}. \tag{14.5}$$

Now let (M,Φ) be a weighted mesh. We consider all possible conformal circle packing metrics, namely, all the possible radius functions $\Gamma : V \to \mathbb{R}^+$. Suppose the vertex set of M is $\{v_1, v_2, \cdots, v_n\}$. Then the corresponding radii are

$$\gamma = (\gamma_1, \gamma_2, \cdots, \gamma_n).$$

For the convenience of discussion, we take the logrithm of the radii,

$$\boldsymbol{u} = (u_1, u_2, \cdots, u_n).$$

All possible $\boldsymbol{u}$'s form an n dimensional space homeomorphic to $\mathbb{R}^n$. We call it the *u-space*. We add one normalization constraint,

$$\sum_{k=1}^{n} u_k = 0.$$

Then all normalized $\boldsymbol{u}$'s form an $(n-1)$ dimensional space homeomorphic to $\mathbb{R}^{n-1}$. We call it *the normalized u-space*.

We define a differential 1-form ω in the u-space,

$$\omega = \sum_{i=1}^{n} K_i du_i.$$

Then

$$d\omega = \sum_{[v_i,v_j]\in M} \left(\frac{\partial K_i}{\partial u_j} - \frac{\partial K_j}{\partial u_i} \right) du_j \wedge du_i = 0,$$

because of equation (14.5), ω is a closed 1-form. Because u-space is simply connected, ω is also exact. We choose a special metric

$$\boldsymbol{u}_0 = (0, 0, \cdots, 0),$$

and define the following energy function

$$E(\boldsymbol{u}) = \int_{\boldsymbol{u}_0}^{\boldsymbol{u}} \sum K_i du_i.$$

Then the energy is independent of the choice of the integration path. The gradient of the energy is

$$\nabla E = (\frac{\partial E}{\partial u_1}, \frac{\partial E}{\partial u_1}, \frac{\partial E}{\partial u_2}, \cdots, \frac{\partial E}{\partial u_n}) = (K_1, K_2, \cdots, K_n).$$

Therefore, discrete surface Ricci flow in equation (14.3) is exactly the negative gradient flow of the energy.

We can further compute the Hessian matrix of E with respect to $\boldsymbol{u}$. If $[v_i, v_j]$ is an edge on the mesh M, then

$$\frac{\partial^2 E(\boldsymbol{u})}{\partial u_i \partial u_j} = \frac{\partial K_i}{\partial u_j} = -w_{ij}.$$

Otherwise $\frac{\partial^2 E(\boldsymbol{u})}{\partial u_i \partial u_j} = 0$. Furthermore

$$\frac{\partial^2 E(\boldsymbol{u})}{\partial u_i^2} = \frac{\partial K_i}{\partial u_i} = \sum_{[v_i,v_j]\in M} w_{ij}.$$

In the normalized u-space, the Hessian matrix is diagonal dominant. Therefore it is positive definite. Hence, the energy $E(\boldsymbol{u})$ has at most one global minimum.

We denote the Hessian matrix as $\Delta(\boldsymbol{u})$. Then from the above calculation, we get

$$\boldsymbol{dk} = \Delta(\boldsymbol{u})\boldsymbol{du}.$$

For the discrete Ricci flow $\boldsymbol{du} = -\boldsymbol{k}dt$, we get

$$\frac{\boldsymbol{dk}}{dt} = -\boldsymbol{k}\Delta(\boldsymbol{u})$$

Therefore, the discrete Gaussian curvature evolves like a heat diffusion process.

In order to compute a metric with the prescribed curvature $\bar{K}$, the energy can be reformulated as

$$E(\boldsymbol{u}) = \int_{\boldsymbol{u}_0}^{\boldsymbol{u}} \sum_{i=1}^{n} (\bar{K}_i - K_i) du_i. \tag{14.6}$$

The desired metric is the global minimizer of the energy. By using Newton's method, the energy can be minimized very efficiently. Fig. 14.8 demonstrates the conformal parameterization results using Newton's method. The parameterization quality highly depends on the choice of the target curvature.

Algorithm 2: Newton's method of discrete Ricci flow

input : A mesh M embedded in $\mathbb{R}^3$, target curvature $\bar{K}$, curvature error threshold ε
output: a circle packing metric (M,Γ,Φ) which induces $\bar{K}$.

Compute the initial circle packing metric (M,Γ_0,ϕ);
Compute the initial curvature K.

$$\boldsymbol{u} \leftarrow 0.$$

while $max|K_i - \bar{K}_i| > \varepsilon$ **do**
  forall *edge* $e = [v_i, v_j] \in M$ **do**
    Compute the edge weight $w_{ij}(\boldsymbol{u})$ to form the Hessien matrix.
  end

$$\begin{aligned} \boldsymbol{du} &\leftarrow H^{-1}(\bar{K} - K) \\ \boldsymbol{u} &\leftarrow \boldsymbol{u} + \boldsymbol{du} \\ \boldsymbol{k} &\leftarrow K(\boldsymbol{u}). \end{aligned}$$

end
$\bar{\boldsymbol{u}} \leftarrow \boldsymbol{u}$

Fig. 14.8. Conformal parameterization using Newton's method for optimizing discrete Ricci energy.

14.5 Isometric Planar Embedding

Once the flat metric is computed, the mesh can be isometrically embedded onto the plane $\mathbb{R}^2$. If the mesh has complicated topology, we need to embed a finite portion of its universal covering space isometrically. In theory, this step is very trivial. But in practice, because of accumulation of the numerical error, planar embedding is very challenging to compute accurately. Instead, we formulate planar embedding as a variational problem, and solve it by using the optimization method. This approach is more robust and accurate than brute force direct embedding.

Suppose M is a topological disk (or a finite portion of the universal covering space of a mesh), with a flat metric $l : E \rightarrow \mathbb{R}^+$. We want to flatten M onto the plane $\mathbb{R}^2$. We denote the map as

$$\phi : M \rightarrow \mathbb{R}^2, \phi(v_i) = (u_i, v_i),$$

such that

$$|\phi(v_i) - \phi(v_j)| = l([v_i, v_j]), \forall [v_i, v_j] \in M.$$

Suppose $f = [v_i, v_j, v_k]$ is a face of the mesh. Then we can embed it in a canonical way in the xy-plane, such that

$$(x_i, y_i) = (0,0), (x_j, y_j) = (l_{ij}, 0), (x_k, y_k) = (l_{ki} \cos\theta_i, l_{ki} \sin\theta_i),$$

where

$$\theta_i = cos^{-1} \frac{l_{ij}^2 + l_{ki}^2 - l_{jk}^2}{2 l_{ij} l_{ki}}.$$

We treat (x_i, y_i) as a local coordinates of $[v_i, v_j, v_k]$. Let

$$\boldsymbol{s}_i = [(x_k, y_k) - (x_j, y_j)] \times \boldsymbol{n}, \boldsymbol{s}_j = [(x_i, y_i) - (x_k, y_k)] \times \boldsymbol{n}, \boldsymbol{s}_k = [(x_j, y_j) - (x_i, y_i)] \times \boldsymbol{n},$$

where $\boldsymbol{n} = (0,0,1)$ is the face normal.

Then $\phi : M \rightarrow \mathbb{R}^2, w_i \rightarrow (u_i, v_i), \forall w_i \in M$, ϕ has two components (ϕ_1, ϕ_2), $\phi_k : M \rightarrow \mathbb{R}$, where ϕ_k's are piecewise linear functions on the mesh. Then

$$\nabla\phi_1(f) = u_i \boldsymbol{s}_i + u_j \boldsymbol{s}_j + u_k \boldsymbol{s}_k, \quad \nabla\phi_2(f) = v_i \boldsymbol{s}_i + v_j \boldsymbol{s}_j + v_k \boldsymbol{s}_k.$$

If ϕ is an isometric map, then

$$\nabla\phi_2(f) = \boldsymbol{n} \times \nabla\phi_1(f).$$

In practice, we fix the image of one face, and minimize the following energy to obtain ϕ,

$$E(\phi) = \sum_{f \in M} |\nabla \phi_2(f) - \boldsymbol{n} \times \nabla \phi_1(f)|^2$$

and

$$\phi = \min_{\tau} E(\tau).$$

The variables of the above energy $E(\phi)$ are the planar images of all vertices $\{(u_i, v_i)\}$, and $E(\phi)$ is quadratic in u_i, v_i. So the optimizing $E(\phi)$ is a linear problem. Fig. 14.9 demonstrates the isometric embedding computed by using this method. The target curvature is set to be zero everywhere on the genus one Kitten model. The conformal flat metric is computed by using Newton's method. Then a fundamental domain is flattened onto the plane isometrically. Fig. 14.10 shows some examples of computing flat metrics of multi-holed annuli. All the boundaries are mapped to circles. Fig. 14.11 shows flat metrics with different boundary curvatures for the same data sets, where two boundaries have zero curvatures everywhere and are mapped to parallel lines.

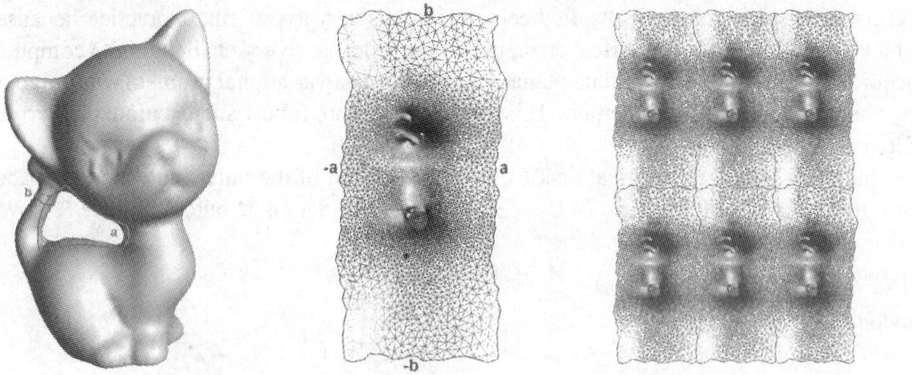

Fig. 14.9. Isometric embedding of the Kitten model with a flat metric.

14.6 Surfaces with Boundaries

Genus zero surfaces with several boundaries can be conformally mapped onto a planar multi-holed annulus. Each boundary is mapped to a circle. If we require the outer boundary to be mapped onto the unit circle, then all such mappings differ by Möbius transformation of the unit disk. If we fix the images of the three points on the outer boundary, then the conformal map is unique. The centers and the radii of the boundary circles are conformal invariants. If two surfaces are conformally equivalent, their planar images are identical.

We first consider a topological disk case. The goal is to find a conformal mapping to map it onto the unit disk. The target curvatures at interior vertices are zeros. The major difficulty is to find the target boundary curvatures. Suppose $\partial M = \{v_1, v_2, \cdots, v_n\}$. If we set the boundary vertex curvatures evenly, $k_i = \frac{2\pi}{n}$, then we won't get a unit disk after running Ricci flow. The final planar shape will be a convex disk. In fact, the target boundary vertex curvatures should satisfy the following requirements,

$$\frac{k_i}{l_{i-1,i}+l_{i,i+1}} \equiv c, \quad \forall v_i \in \partial M,$$

and $\sum_{v_i \in \partial M} k_i = 2\pi$, where l_{ij} are the target edge lengths, k_i is the target vertex curvature. At the beginning of the algorithm, we don't know the target edge length. Therefore, we can not set the target curvature. In the algorithm, we need to adjust the target curvature during the computation.

Algorithm 3: Newton's method of discrete Ricci flow for multi-holed annulus

input : A multi-holed annulus M
output: a flat circle packing metric (M, Γ, Φ) which maps the boundaries to circles.

Compute the initial circle packing metric (M, Γ_0, ϕ);
Compute the initial curvature K.;
Trace boundaries, store each boundary as a list of ordered vertices

$$\Gamma_1, \Gamma_2, \cdots, \Gamma_m.$$

Set Γ_1 as the outer boundary,

$$\theta_1 \leftarrow 2\pi, \theta_2 \leftarrow -2\pi, \cdots, \theta_m \leftarrow -2\pi.$$

forall *Interior Vertex* $v \in M$ **do**

$$\bar{K}(v) \leftarrow 0.$$

end
forall *Boundaries* Γ_k **do**
  forall $v \in \Gamma_i$ **do**

$$\bar{k}(v) \leftarrow \frac{\theta_k}{|\Gamma_k|}$$

  end
end
repeat
  Compute target metric according to $\bar{K}$ using Newton's method in algorithm 2;
  forall Boundaries Γ_k **do**
    Compute the total length of Γ_k under current metric, s_k;
  end
  forall $v_i \in \Gamma_i$ **do**

$$\bar{k}(v_i) \leftarrow \theta_k \frac{l_{i-1,i}+l_{i,i+1}}{2s_k}$$

  end
until The maximal difference of current target curvature and previous target curvature is less than ε;

14.7 Optimal Parameterization Using Ricci flow

Suppose M is a topological disk. Then for an arbitrary planar domain D, there will be conformal mappings from M to the D. Although conformal mappings introduce no angle distortion, they introduce area distortions. In engineering practice, it is highly desirable to

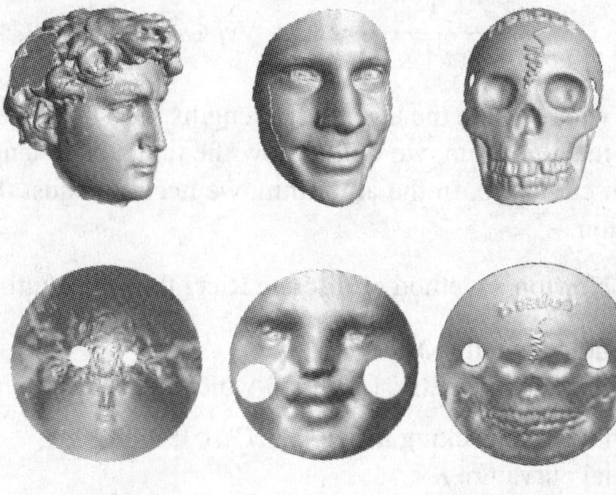

Fig. 14.10. Surfaces which are topologically equivalent but not conformal equivalent.

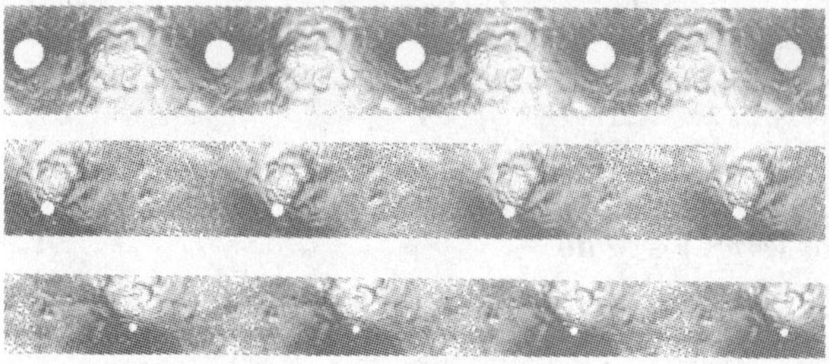

Fig. 14.11. Tiling of different global conformal parameterizations for David head surface with three boundaries using curvature control by discrete Euclidean Ricci Flow.

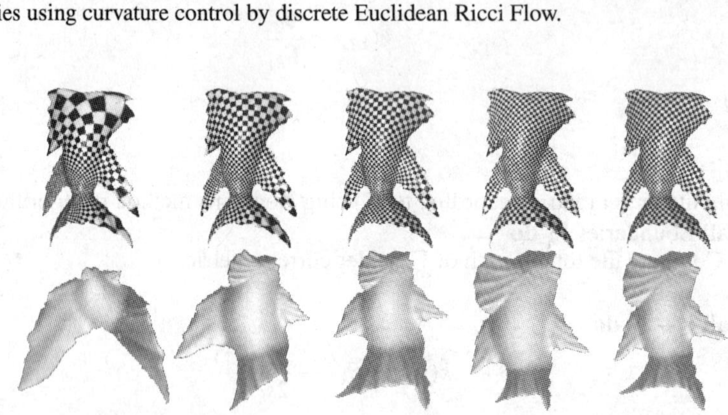

Fig. 14.12. Conformal parameterization optimization process. The initial parameterization is gradually optimized to reach the most uniform result.

obtain conformal mappings with uniform area distortion. This section discusses one rigorous method to choose the target domain to achieve the most uniform conformal mappings.

Given a weighted mesh (M,Φ) with n vertices, all the possible curvatures form a $n-1$ dimensional convex polytope determined by both the connectivity of M and the edge angle Φ, denoted as Ω_k. For any interior vertex, the target curvature will be zero. According to the discrete Gauss-Bonnet theorem, the summation of boundary vertex curvature equals to $2\pi\chi(M)$. All these curvatures form a convex polytope as defined in the following:

Definition 14.13 (Admissible curvature space). *Given a weighted mesh (M,Φ) with boundaries, the admissible curvature space is the intersection of hyper-planes with Ω_k,*

$$\Pi_M := \bigcap_{v_i \notin \partial M} \{k_i = 0\} \bigcap \{ \sum_{v_j \in \partial M} k_j = 2\pi\chi(M)\} \bigcap \Omega_k.$$

Suppose $\boldsymbol{u}$ is a circle packing metric on (M,Φ) in the normalized $\boldsymbol{u}$-space. Then $\boldsymbol{u}$ uniquely determines a curvature $\boldsymbol{k}$. From the previous discussion, we can compute the Jacobi matrix of the mapping

$$K : \boldsymbol{u} \to \boldsymbol{k}.$$

By an explicit calculation, we can get

$$d\boldsymbol{k} = \Delta d\boldsymbol{u},$$

where Δ is the Hessian matrix of the discrete Ricci energy (14.6). In the normalized u-space, Δ is positive definite. Therefore it is invertible, $d\boldsymbol{u} = \Delta^{-1} d\boldsymbol{k}$.

The optimal parameterization problem is equivalent to optimizing some energy form $E(\boldsymbol{k})$, which measures the uniformity of the area distortion in the admissible curvature space.

Problem 14.14 (Optimal parameterization problem). Compute the minima of the energy $E(\boldsymbol{k})$ in the admissible curvature space Π_M.

If the energy is differentiable, the problem can be solved by using *projected gradient* method. We compute the gradient of $E(\boldsymbol{k})$ with respect to $\boldsymbol{k}$, denoted as ∇E. Then project the gradient to the affine subspace Π_k, and update $\boldsymbol{k}$ along the projected gradient direction, until the projected gradient equals to zero. Namely, at a critical point of $E(\boldsymbol{k})$, $\nabla_k E$ is orthogonal to the admissible curvature space.

Let $\boldsymbol{e}_i$ denote the normal to the plane $\{k_i = 0\}$, d be the unit normal to the plane $\{\sum_{v_i \in \partial M} k_i = 2\pi\chi(M)\}$. The following algorithm optimizes $E(\boldsymbol{k})$ by using the projected gradient method.

Algorithm 4: Optimal discrete conformal prameterization using projected gradient method

input : A mesh M with boundaries
output: a flat circle packing metric (M,Γ,Φ) which minimizes an energy $E(\boldsymbol{k})$ or $E(\boldsymbol{u})$.

Compute the initial circle packing metric (M,Γ_0,ϕ);
Compute the initial curvature K.;
repeat
  $\boldsymbol{u} \leftarrow K^{-1}(\boldsymbol{k})$ Compute the gradient of E with respect to $\boldsymbol{u}$, $\nabla_u E$;
  $\nabla_k E \leftarrow \Delta(\boldsymbol{u})^{-1} \nabla_u E$;
  forall $v_i \notin \partial M$ **do**

$$\nabla_k E \leftarrow \nabla_k E - \langle \nabla_k E, \boldsymbol{e}_i \rangle \boldsymbol{e}_i$$

  end
  $\nabla_k E \leftarrow \nabla_k E - \langle \nabla_k E, \boldsymbol{d} \rangle \boldsymbol{d}$;
  $\boldsymbol{k} \leftarrow \boldsymbol{k} - \lambda \nabla_k E$;
until $|\nabla_k E| < \varepsilon$;

Proposition 14.15. *Suppose $\mathbf{k}$ is an interior point of Π_k, also an optimum for an energy form $E(\mathbf{u})$. Then all components of $\nabla_k E$ corresponding to the boundary vertices are equal.*

Proof. If $\mathbf{k}$ is an optima of $E(\mathbf{k})$, then $\nabla_k E \perp \Pi_M$. Therefore

$$\nabla_k E = \sum_{v_i \notin \partial M} \lambda_i \mathbf{e}_i + \mu \mathbf{d}.$$

We can design different energy forms $E(\mathbf{k})$ to measure the uniformity of the conformal parameterization. Because $\mathbf{u}$ is in the normalized u-space,

$$\sum_{v_i \in M} k_i = 0,$$

the variance

$$E(\mathbf{u}) = \sum_{v_i \in M} (u_i - u_i^0)^2$$

measure the uniformity of the parameterization, where $\mathbf{u}^0$ is the initial circle packing metric. This energy can be reformulated as

$$\sum_{v_i \in M} (u_i - u_i^0)^2 = \int_{\mathbf{u}_0}^{\mathbf{u}} (\mathbf{u} - \mathbf{u}^0) d\mathbf{u} = \int_{\mathbf{k}_0}^{\mathbf{k}} (\mathbf{u} - \mathbf{u}^0) \Delta^{-1}(\mathbf{k}) d\mathbf{k}.$$

Because Δ is positive definite, therefore Δ^{-1} is also positive definite. Then we can replace Δ^{-1} by other positive definite matrix, and expect that the new energy form also measures the uniformity of the parameterization.

We choose the Legendre dual of the Ricci energy,

$$E(\mathbf{k}) = \int_{\mathbf{k}_0}^{\mathbf{k}} \sum_i (u_i - u_i^0) dk_i.$$

The Hessien matrix of $E(\mathbf{k})$ is $\Delta^{-1}(\mathbf{k})$, which is positive definite when restricted to the normalized u-space. Therefore the energy is convex. It has at most one global minimum, which can be achieved by direct projected gradient method. It can also be minimized by using the following curvature diffusion with free boundaries,

Algorithm 5: Curvature Diffusion with Free Boundaries

input : A mesh M with boundaries
output: A most uniform circle packing metric
Compute the initial circle packing metric (M, Φ, Γ);
Compute the initial curvature $\mathbf{k}_0$.
while $max|k_i| > \varepsilon$ **do**
  forall $v_i \notin \partial M$ **do**
    $du_i \leftarrow -k_i$;
    $u_i \leftarrow u_i + \lambda du_i$;
  end
  $c \leftarrow \frac{\sum_{v_i \in M} u_i}{|V|}$ **forall** *vertex* $v_i \in M$ **do**
    $u_i \leftarrow u_i - c$.
  end
end

This algorithm can also lead to the global minimum of the uniformity energy. Fig. 14.12, 14.13 show two cases of optimal conformal parameterization by using projected gradient method. The initial parameterizations are highly non-uniform. The final results are much more improved.

Fig. 14.13. Optimized conformal parameterization of horse and camel models. The initial parameterizations on the left are gradually improved and reached the most uniform conformal mapping on the right.

15

Hyperbolic Ricci Flow

In the previous discussion, we assumed the surface was embedded in the three dimensional Euclidean space $\mathbb{R}^3$. Therefore the length and the angles were computed by using Euclidean geometry. Specifically, we use a piecewise-linear triangle mesh to approximate the surface. The mesh is composed by gluing many triangles coherently, and each triangle is a Euclidean triangle. The corner angles on the triangle can be calculated by using edge lengths by the Euclidean cosine law. In this case, we say the computation is using *the Euclidean background geometry*.

We generalize the method to *the hyperbolic background geometry* in the following way. Given a mesh M, we treat each face as a hyperbolic triangle, each edge is a geodesic (a hyperbolic line), the corner angles and the edge lengths are determined by the hyperbolic cosine law.

Recall that the cosine law for triangles with different background geometries are

$$\begin{array}{lll} l_k^2 & = l_i^2 + l_j^2 - 2 l_i l_j \cos\theta_k & \mathbb{E}^2 \\ \cosh l_k^2 & = \cosh l_i \cosh l_j + \sinh l_i \sinh l_j \cos\theta_k & \mathbb{H}^2 \\ \cos l_k^2 & = \cos l_i \cos l_j - \sin l_i \sin l_j \cos\theta_k & \mathbb{S}^2 \end{array}$$

If a mesh M has a circle packing metric in hyperbolic background geometry, the edge length should be calculated by using hyperbolic cosine law. In detail, suppose vertices v_i and v_j are connected by an edge $[v_i, v_j]$. The circles centered at v_i and v_j have radii γ_i, γ_j respectively. Then the edge length of $[v_i, v_j]$ is

$$\cosh l_{ij} = \cosh\gamma_i \cosh\gamma_j + \sinh\gamma_i \sinh\gamma_j \cos\phi_{ij}.$$

The discrete Gaussian curvature of each vertex is calculated in a similar way as the Euclidean Ricci flow,

$$K_i = \begin{cases} 2\pi - \sum_{f_{ijk}\in M} \theta_i^{jk} & \text{interior vertex} \\ \pi - \sum_{f_{ijk}\in M} \theta_i^{jk} & \text{boundary vertex.} \end{cases}$$

The discrete conformal factor u_i is defined differently for a hyperbolic or a spherical mesh M with a circle packing metric (M, Γ, Φ),

$$u_i = \begin{cases} \log\gamma_i & \mathbb{E}^2 \\ \log\tanh\frac{\gamma_i}{2} & \mathbb{H}^2 \\ \log\tan\frac{\gamma_i}{2} & \mathbb{S}^2 \end{cases}$$

The discrete Ricci flow for hyperbolic or spherical mesh is in the same form:

$$\frac{du_i(t)}{dt} = \bar{K}_i - K_i.$$

In practice, we always set the target curvature $\bar{K}_i$ be zero at all vertices. The discrete Ricci flow is the negative gradient flow of the discrete Ricci energy, which has the same form as the discrete Euclidean Ricci flow. Suppose the vertex set is $\{v_1, v_2, \cdots, v_n\}$, the discrete conformal factor vector is $\boldsymbol{u} = \{u_1, u_2, \cdots, u_n\}$. Then the following symmetry holds for hyperbolic or spherical meshes

$$\frac{\partial K_i}{\partial u_j} = \frac{\partial K_j}{\partial u_i}.$$

Therefore the discrete hyperbolic, Euclidean or spherical energy is

$$E(\boldsymbol{u}) = \int_{\boldsymbol{u}_0}^{\boldsymbol{u}} \sum_{i=1}^{n} (\bar{K}_i - K_i) du_i.$$

The hyperbolic Ricci energy is convex if all the intersection angles ϕ_{ij} are acute. Therefore, it has a global minimum, which is the metric $\bar{\boldsymbol{u}}$. The desired curvature of $\bar{\boldsymbol{u}}$ is curvature $\bar{\boldsymbol{k}}$. The spherical Ricci energy is not convex.

In order to get a hyperbolic metric for a surface with negative Euler number, we can simply set the target curvatures to zeros at all vertices, and minimize the discrete hyperbolic Ricci energy. Because the energy is convex, Newton's method leads to the unique global minimum with arbitrary starting point of $\boldsymbol{u}$ stably. The elements of the Hessian matrix are computed by using the following formulae:

$$\frac{\partial^2 E(\boldsymbol{u})}{\partial u_j \partial u_i} = -\frac{K_i}{\partial u_j}.$$

Details can be deduced by direct computation as shown in [JLYG].

15.1 Hyperbolic Embedding

Suppose M is a mesh with a circle packing metric in hyperbolic background geometry. Let $(\bar{M}, \pi)$ be the universal covering space of M, where π is the projection. Suppose (Γ, Φ) is a hyperbolic circle packing metric on M, then the pull back metric on $\bar{M}$ is also a hyperbolic circle packing metric, denoted as $(\bar{M}, \pi^*\Gamma, \pi^*\Phi)$. If all the discrete vertex curvatures of (M, Γ, Φ) are zeros. Then all the discrete vertex curvatures of $(\bar{M}, \pi^*\Gamma, \pi^*\Phi)$ are also zeros. Therefore,$(\bar{M}, \pi^*\Gamma, \pi^*\Phi)$ can be isometrically embedded onto the hyperbolic space $\mathbb{H}^2$. The embedding process is more complicated than the Euclidean case.

15.1.1 Embedding One Face

The most popular model for the hyperbolic space $\mathbb{H}^2$ is the Poincaré disk model. See Chapter 3. The Poincaré disk model is the unit disk in the complex plane,

$$|z| < 1,$$

with the Riemannian metric:

$$ds^2 = \frac{4dzd\bar{z}}{(1-\bar{z}z)^2}, \quad z \in \mathbb{C}.$$

This metric is conformal to the Euclidean metric $dzd\bar{z}$. All the rigid motions are Möbius transformations:

$$z \to e^{i\theta}\frac{z-z_0}{1-\bar{z}_0 z}, \quad |z_0|<1, \quad \theta \in [0,2\pi).$$

Suppose $p,q \in \mathbb{H}^2$ are two points in the Poincaré disk. The hyperbolic geodesic (hyperbolic line) is the circular arc through p and q and orthogonal to the unit circle $|z|=1$. The hyperbolic line intersects the unit circle at a and b, as shown in Fig. 15.1. The hyperbolic distance between p and q is

$$d(p,q) = \left|\log \frac{|p-a|/|p-b|}{|q-a|/|q-b|}\right|.$$

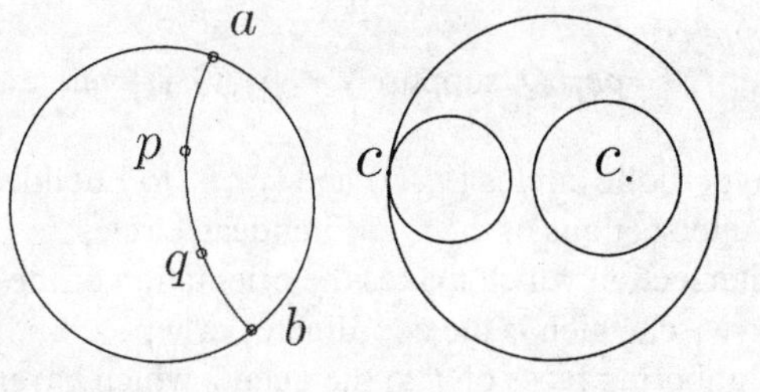

Fig. 15.1. Hyperbolic line and hyperbolic circle.

The hyperbolic lines through the origin are also Euclidean lines. Suppose we embed a hyperbolic triangle as shown in the middle of the Fig. 15.2 with

$$v_i = 0, v_j = \frac{e^{l_k}-1}{e^{l_k}+1}, v_k = \frac{e^{l_j}-1}{e^{l_j}+1}e^{i\theta_i}. \tag{15.1}$$

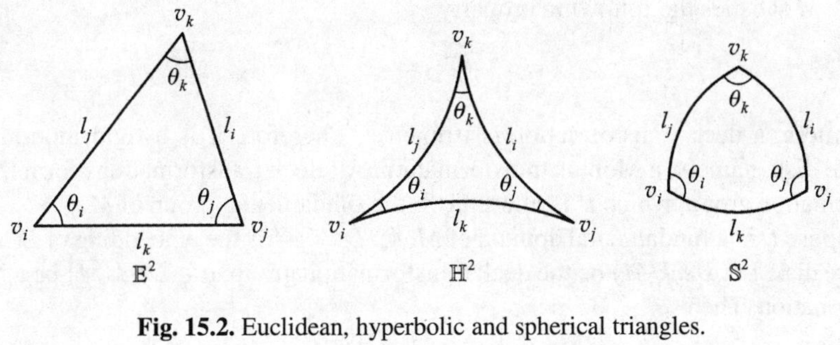

Fig. 15.2. Euclidean, hyperbolic and spherical triangles.

In the general case, suppose v_i, v_j have been set. We need to determine v_k. We use (c,γ) to denote a hyperbolic circle centered at c with radius γ. Then v_k is one of the two intersection points of (v_i,l_j) and (v_j,l_i), which makes the orientation of the triangle $[v_i,v_j,v_k]$ counter clock wise.

In the Poincaré disk, the locus of a hyperbolic circle (c,γ) is also a Euclidean circle (C,R), where

$$C = \frac{2-2\mu^2}{1-\mu^2|c|^2}c, \quad R^2 = |C|^2 - \frac{|c|^2-\mu^2}{1-\mu^2|c|^2},$$

and

$$\mu = \frac{e^r-1}{e^r+1}.$$

Therefore, the center of a hyperbolic circle may not be the center of the circle in the Euclidean sense. It is possible that a hyperbolic circle is with an infinite radius. In this case, the center is on the unit circle, which is the tangent point between the hyperbolic circle and the unit circle as shown in the right frame of Fig. 15.1. The left hyperbolic circle has infinite radius.

The following algorithm will describe how to embed a hyperbolic mesh, which is a topological disk, onto the Poincaré disk.

Algorithm 6: Hyperbolic embedding of a topological disk.

input : A simply connected mesh M with single boundary and hyperbolic metric
output: Hyperbolic embedding of M onto the Poincaré disk

Select a root face f, embed f using formula 15.1;
Put all faces sharing an edge with f in a queue Q;
while *Q is not empty* **do**
    Pop the queue, $f \leftarrow$ *pop* Q, suppose $f = [v_i, v_j, v_k]$ where v_i, v_j have been embedded;
    Convert the hyperbolic circles (v_i, l_j) and (v_j, l_i) to Euclidean circles.;
    Compute the intersections of the two Euclidean circles;
    Choose the intersection which makes the orientation of the face counter-clock-wise, which is the coordinates of v_k;
    Put all the neighboring faces of f to the queue, which haven't been embedded;
end

15.1.2 Hyperbolic Embedding of the Universal Covering Space

The whole universal covering space $(\bar{M}, \pi)$ with the hyperbolic metric of a high genus surface M can be isometrically embedded onto the Poincaré disk. A deck transformation $\rho : \bar{M} \to \bar{M}$ satisfies the following property:

$$\pi \circ \rho = \pi.$$

Furthermore, a deck transformation is isometric. Therefore it is a rigid motion of the Poincaré disk, namely, a Möbius transformation. All deck transformations form the deck transformation group, which is isomorphic to the fundamental group of M.

Suppose D is a fundamental domain of M, $\phi : D \to \mathbb{H}^2$ is the embedding of D onto the Poincaré disk. Let $Deck(M)$ be the deck transformation group, $\rho \in Deck(M)$ be a Möbius transformation. Then

$$\bigcup_{\rho \in Deck(M)} \rho \circ \phi(D) = \mathbb{H}^2,$$

i.e. all copies of the embedding of a fundamental domain covers the whole Poincaré disk. This shows the embedding of the universal covering space covers the whole Poincaré disk.

The key is to find the deck group generators. This can be found in the following manner. Suppose M is a closed genus g mesh. First we choose a base point $v \in M$. Then we choose a canonical fundamental group generators

$$\{a_1, b_1, a_2, b_2, \cdots, a_g, b_g\}.$$

We slice the mesh along the fundamental group generators to form a fundamental domain D. The boundary of D is

$$\partial D = a_1 b_1 a_1^{-1} b_1^{-1} a_2 b_2 a_2^{-1} b_2^{-1} \cdots a_g b_g a_g^{-1} b_g^{-1}.$$

Then we embed D onto the Poincaré disk by using the method described in the last subsection. $\phi : D \to \mathbb{H}^2$ is an embedding.

For simplicity, we omit the embedding ϕ in the following discussion, and assume D has been embedded in $\mathbb{H}^2$ already. Then a_i, b_j are curves in the $\mathbb{H}^2$. There exists a unique Möbius transformation ρ_k, which maps a_k to a_k^{-1}, namely for $p \in a_k$, $q \in a_k^{-1}$, $\pi(p) = \pi(q)$, then

$$\rho(p) = q, \quad \rho(a_k) = a_k^{-1}.$$

Similarly, there exists a unique τ_k which maps b_k to b_k^{-1}. Then

$$\{\rho_1, \tau_1, \rho_2, \tau_2, \cdots, \rho_g, \tau_g\}$$

are the generators of the deck transformation group of M (Fuchsian group generator).

Now, let's focus on the computation of ρ_1. The other generators can be obtained in the same way. Suppose the starting and ending points of a_1 are s, t, $\partial a_1 = t - s$. The starting and the ending point of a_1^{-1} is t_1, s_1, $\partial a_1^{-1} = s_1 - t_1$. It is obvious that all of them are the pre-images of the base point, $\pi : \{s, t, s_1, t_1\} \to v$. We want to find a Möbius transformation ρ_1, such that

$$\rho_1(s) = s_1, \rho_1(t) = t_1.$$

Suppose the hyperbolic line through s and t is $\overline{st}$ and the hyperbolic line through s_1 and t_1 is $\overline{s_1 t_1}$. Then ρ_1 maps $\overline{st}$ to $\overline{s_1 t_1}$. We first construct a Möbius transformation sending s to the origin,

$$\phi_1 : z \to \frac{z - s}{1 - \bar{s}z}.$$

Then ϕ maps $\overline{st}$ to a radial Euclidean line. Suppose the angle from the real axis to $\phi(\overline{st})$ is θ, and

$$\phi_2 : z \to e^{i\theta}.$$

Then $\phi_2 \circ \phi_1$ maps $\overline{st}$ to the real axis, s to the origin, t to the point $\frac{e^d - 1}{e^d + 1}$, where d is the hyperbolic distance between s and t. We can use similar method to construct a Möbius transformation $\bar{\phi}_2 \circ \bar{\phi}_1$ sending s_1 to the origin, t_1 to $\frac{e^d - 1}{e^d + 1}$. Then

$$\rho_1 = \bar{\phi}_1^{-1} \circ \bar{\phi}_2^{-1} \phi_2 \circ \phi_1$$

maps a_1 to a_1^{-1}.

Once we have the generators of the Fuchsian group, we can transform the embedding of the fundamental domain and glue them coherently. Suppose D is the fundamental domain, $p \in D$ is an interior point, D_k is another fundamental domain and p_k is an interior point, $\pi(p) = \pi(p_k)$. Suppose $\gamma \subset \bar{M}$ is a path connecting p to p_k, then $\pi(\gamma)$ is a loop in M. Suppose the homotopy class of $\pi(\gamma)$ is

$$[\gamma] = s_1 s_2 \cdots s_k,$$

where the symbol

$$s_i \in \{a_1, b_1, a_1^{-1}, b_1^{-1}, \cdots, a_g, b_g, a_g^{-1}, b_g^{-1}\}, \quad 1 \leqslant i \leqslant k.$$

The corresponding Möbius transformation can be computed in the following way by replacing the generators of fundamental group by the generators in Fuchsian group:

1. Replace a_i by τ_i^{-1},
2. Replace b_i by ρ_i.

Then we get a unique Möbius transformation, which maps D to D_k.

Algorithm 7: Hyperbolic embedding of the universal covering space of a high genus mesh

input : A high genus closed mesh M with hyperbolic metric
output: Hyperbolic embedding of the finite portion of the universal covering space of M onto the Poincaré disk

Compute a canonical fundamental group generators;

$$\{a_1, b_1, a_2, b_2, \cdots, a_g, b_g\}.$$

Slice M along the generators to form a fundamental domain D;
Embed D onto the Poincaré disk using the algorithm 6;
Compute Möbius transformations,

$$\{\rho_1, \tau_1, \rho_2, \tau_2, \cdots, \rho_g, \tau_g\}$$

ρ_k maps a_k to a_k^{-1}, τ_k maps b_k to b_k^{-1}.;
For any element in the fundamental group γ_k, compute the corresponding deck transformation by replacing the generators of the fundamental group by the generators of the Fuchsian group;
For each element τ in Fuchsian group, map D to $\tau(D)$;
Stich the $\tau(D)$'s coherently along their boundaries.

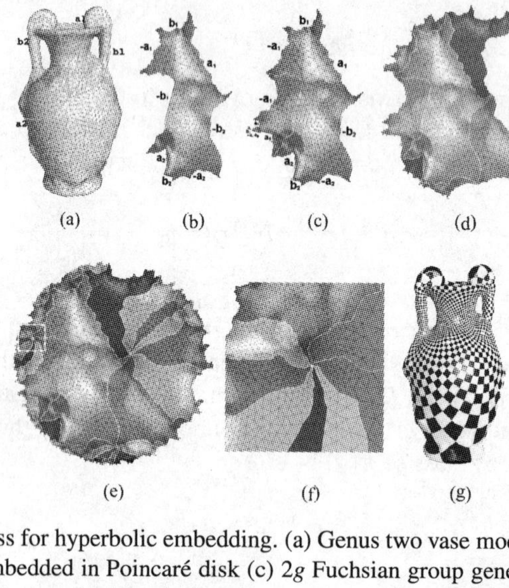

Fig. 15.3. The process for hyperbolic embedding. (a) Genus two vase model (b) One canonical fundamental domain embedded in Poincaré disk (c) $2g$ Fuchsian group generators act on vase model, which are rigid motions in the hyperbolic space. Different color indicates different periods (fundamental domains). The generators maps the central period to the colored ones respectively. (d) Different fundamental domains are glued together by the Fuchsian transformation. (e) All fundamental domains adjacent to the first one are glued together. (f) A local region in (e) is zoomed. All patches are coherently glued together, which demonstrates the accuracy of the algorithm. (g) The mapping is conformal illustrated by a checker-board texturemapping.

Once the universal covering space is embedded onto the Poincaré disk, we can use hyperbolic lines to separate the fundamental domains. Each fundamental domain becomes a hyperbolic polygon, which is called the fundamental polygon. Fig. 15.3 illustrates the whole process of computing hyperbolic embedding. Fig. 15.4 demonstrates the hyperbolic embedding of several high genus surfaces.

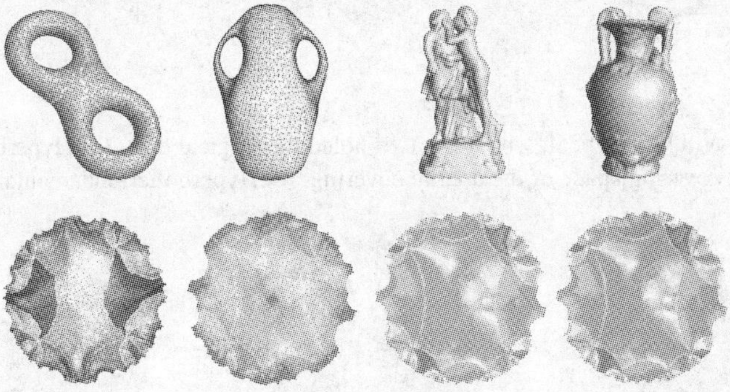

Fig. 15.4. Hyperbolic embedding of the universal covering spaces of high genus surfaces.

15.2 Surfaces with Boundaries

Suppose M is a compact surface with boundaries. If Euler number $\chi(M)$ of M is negative, then we can compute its hyperbolic metric by double covering. We denote the double covering of M as $\bar{M}$. Then $\bar{M}$ is a symmetric closed surface, so that $\chi(\bar{M}) = 2\chi(M)$ is still negative. We compute the hyperbolic metric of $\bar{M}$. Then all boundaries become hyperbolic lines. This process constructs the hyperbolic metric of M.

Figure 15.5 illustrates the process. The David head model is a two-holed annulus, its double covering is a genus two torus. Two copies of the original surface on the doubled one are illustrated by different shading. We compute its hyperbolic metric by optimizing the hyperbolic Ricci energy and embed a finite portion of its universal covering space onto the Poincaré disk. We can see that the boundaries of the original surface are mapped to hyperbolic lines.

The hyperbolic space $\mathbb{H}^2$ can also be modeled by using upper half plane model $\{z : \operatorname{Im} z > 0\}$. The Riemannian metric is defined as

$$ds^2 = \frac{dzd\bar{z}}{y^2}.$$

The geodesics are circular arcs or lines perpendicular to the real axis. The following conformal map converts the Poincaré disk to the upper half plane,

$$z \to i\frac{1+z}{1-z}. \tag{15.2}$$

In Fig. 15.5 and 15.6, we compute the fundamental polygon in the Poincaré disk first. Then convert the $\mathbb{H}^2$ model from the Poincaré disk to the upper half plane. The fundamental polygons in the upper half plane are shown in the third frames respectively.

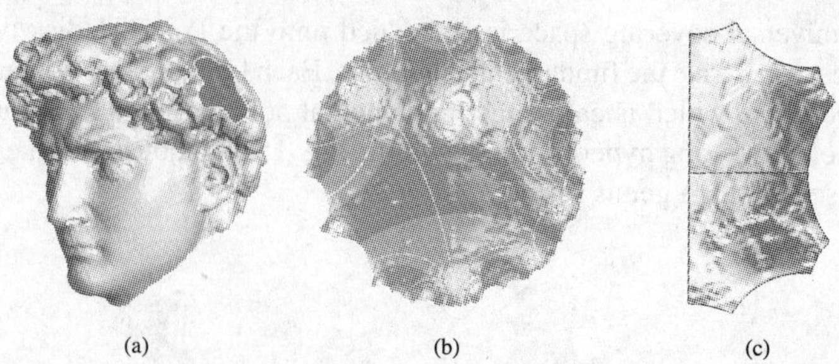

Fig. 15.5. (a) David head model, which is a two-holed topological disk. (b) Hyperbolic embedding of the universal covering space of the double covering. (c) Hyperbolic fundamental polygon.

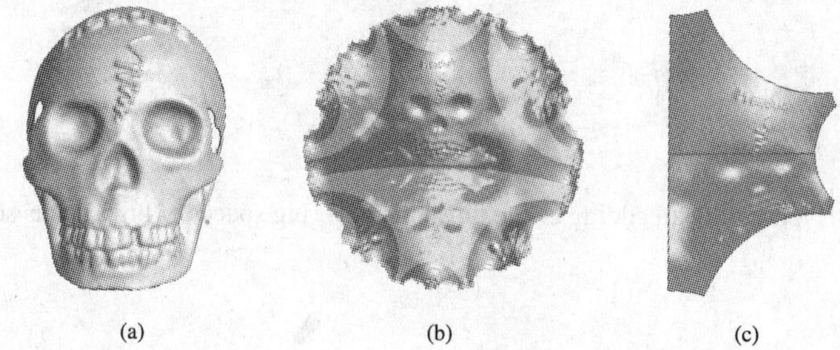

Fig. 15.6. (a) Skull model, which is a two-holed topological disk. (b) Hyperbolic embedding of the universal covering space of the double covering. (c) Hyperbolic fundamental polygon.

We also call a two-holed annulus as a pair of pants. Given a pair of pants with Riemannian metric, it has a unique hyperbolic metric which is conformal to its original metric. Three boundaries are geodesics under the hyperbolic metric. (Compare Lemma 9.4). Suppose the boundaries are $\gamma_1, \gamma_2, \gamma_3$. The hyperbolic lengths of them are (l_1, l_2, l_3). Then the triple (l_1, l_2, l_3) represents the conformal class of the surface. The triple is the Teichmüller space coordinates of the surface. In details, given two surfaces M_1 and M_2, which are two pairs of pants and with hyperbolic metrics, suppose $\partial M_1 = \{\gamma_1^1, \gamma_2^1, \gamma_3^1\}$, $\partial M_2 = \{\gamma_1^2, \gamma_2^2, \gamma_3^2\}$, the length of γ_1^k equals to that of $\gamma_2^k, k = 1,2,3$. Then there exists a conformal map $\phi : M_1 \to M_2$, such that $\phi(\gamma_1^k) = \gamma_2^k$.

Suppose M is a pair of pants with a hyperbolic metric, all boundaries $\{\gamma_1, \gamma_2, \gamma_3\}$ are geodesics. Then there exist unique geodesics $\{\tau_1, \tau_2, \tau_3\}$, such that

$$\tau_i \perp \gamma_j, \tau_i \perp \gamma_k, i \neq j, i \neq k.$$

These geodesics bisect the boundaries and separate the pair of pants to two isometric hyperbolic hexagons, whose inner angles are right angles. Fig. 15.7 shows the two hyperbolic hexagons of the fundamental polygon of the David head model in Fig. 15.5. The center circle is the boundary of the Poincaré disk. $\{\gamma_1, \gamma_2, \gamma_3\}$ are the boundaries of the surface, which are hyperbolic lines, $\{\tau_1, \tau_2, \tau_3\}$ are the hyperbolic lines orthogonal to the boundaries. The surface is partitioned into two identical hyperbolic hexagons with right corner angles.

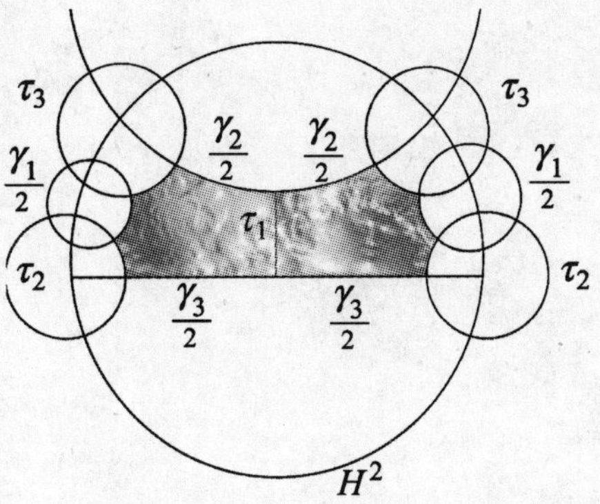

Fig. 15.7. Hyperbolic embedding of a pair of pants.

Further Readings

The thorough textbook on Ricci flow theories is [CLN]. The surface Ricci flow work was introduced by Hamilton in [Ham]. The connection between Ricci flow and circle packing method is analyzed in [CL]. The circle packing was proposed by Thurston in [Thu]. An excellent text book on circle packing is by Stephenson [St]. The circle pattern was proposed in [BS]. It is applied for surface parameterization in [KSS]. The discrete hyperbolic Ricci flow for computing general geometric structures is first introduced in [JLG]. The discrete Euclidean Ricci flow has been applied for computing manifold splines with single singularity in [HJY]. The discrete Ricci flow for computing geodesic spectrum as the coordinates in the shape space is introduced in [JLYG]. More references on parameterization algorithms can be found in the surveys [FH] and [SPR].

References

[Cha] Chavel, Isaac. Isoperimetric inequalities: Differential geometric and analytic perspectives. Cambridge Tracts in Mathematics, 145. Cambridge University Press, Cambridge, 2001. xii+268 pp. ISBN: 0-521-80267-9 (Reviewer: Thierry Coulhon)

[Gru] Grünbaum, Branko. Convex polytopes. Second edition. Prepared and with a preface by Volker Kaibel, Victor Klee and Günter M. Ziegler. Graduate Texts in Mathematics, 221. Springer-Verlag, New York, 2003. xvi+468 pp. ISBN: 0-387-00424-6; 0-387-40409-0 (Reviewer: Alexander Zvonkin)

[Thu] Thurston, William P. Three-dimensional geometry and topology. Vol. 1. Edited by Silvio Levy. Princeton Mathematical Series, 35. Princeton University Press, Princeton, NJ, 1997. x+311 pp. ISBN: 0-691-08304-5 (Reviewer: Athanase Papadopoulos)

[An] Andreev, E. M., Convex polyhedra in Lobachevsky spaces. (Russian) Mat. Sb. (N.S.) 81 (123) 1970, 445–478.

[Le] Leibon, Gregory. Characterizing the Delaunay decompositions of compact hyperbolic surfaces. Geom. Topol. 6 (2002), 361–391.

[Lu1] Luo, Feng. Rigidity of polyhedral surfaces. arXiv:math.GT/0612714.

[Lu2] Luo, Feng. A characterization of spherical polyhedron surfaces. J. Differential Geom. 74, (2006), no. 3, 407–424.

[Lu3] Luo, Feng. On Teichmller spaces of surfaces with boundary, Duke Mathematical Journal. Volume 139, Number 3 (2007), 463–482.

[Ri] Rivin, Igor. Euclidean structures on simplicial surfaces and hyperbolic volume, Ann. of Math. (2) 139 (1994), no. 3, 553–580.

[Gu1] Guo, Ren. On parameterizations of Teichmuller spaces of surfaces with boundary. arXiv: math.GT/0612221.

[Gu2] Guo, Ren. Geometric angle structures on triangulated surfaces. arXiv:math.GT/0601486.

[FC] Vladimir V. Fok, Leonid O. Chekhov. Quantum Teichmuller spaces. (Russian) Teoret. Mat. Fiz. 120 (1999), 511528; translation in Theoret. and Math. Phys. 120 (1999), 1245–1259.

[Ka] Kashaev Rinat. Quantization of Teichmuller spaces and the quantum dilogarithm, Lett. Math. Phys. 43 (1998), 105–115.

[BL] Bonahon, F.; Liu, X. Representations of the quantum Teichmuller space and invariants of surface diffeomorphisms, arXiv:math.GT/0407086.

[Te] Teschner, J. An analog of a modular functor from quantized Teichmuller theory, arXiv:math.QA/0510174.

[Us] Ushijima, Akira. A canonical cellular decomposition of the Teichmller space of compact surfaces with boundary. Comm. Math. Phys. 201 (1999), no. 2, 305–326.

[Ko] Kojima, Sadayoshi. Polyhedral decomposition of hyperbolic 3-manifolds with totally geodesic boundary. Aspects of low-dimensional manifolds, 93–112, Adv. Stud. Pure Math., 20, Kinokuniya, Tokyo, 1992.

[CV1] Colin de Verdiere, Yves. Un principe variationnel pour les empilements de cercles, Invent. Math. 104 (1991), no. 3, 655–669.

[MaR] Marden, Al; Rodin, Burt. On Thurstons formulation and proof of Andreevs theorem. Computational methods and function theory (Valparalo, 1989), 103–115, Lecture Notes in Math., 1435, Springer, Berlin, 1990.

[Br] Brägger, W. Kreispackungen und Triangulierungen. Enseign. Math., 38:200–217,1992.

[CKP] Cohn, Henry; Kenyon, Richard; Propp, James. A variational principle for domino tiling. J. Amer. Math. Soc. 14 (2001), no. 2, 297–346.

[CV2] Colin de Verdiere, Yves. private communication.

[BS] Bobenko, Alexander I.; Springborn, Boris A. Variational principles for circle patterns and Koebes theorem. Trans. Amer. Math. Soc. 356 (2004), no. 2, 659–689.

[Sch2] Schlenker, Jean-Marc Circle patterns on singular surfaces, arXiv:math.DG/0601531.

[Spr] Boris A. Springborn, A variational principle for weighted Delaunay triangulations and hyperideal polyhedra, arXiv: math/0603097.

[Cro] Cromwell; Peter R. Polyhedra. One of the most charming chapters of geometry. Cambridge University Press, Cambridge, 1997. xiv+451 pp. ISBN: 0-521-55432-2

[HR] Hodgson, Craig D.; Rivin, Igor. A characterization of compact convex polyhedra in hyperbolic 3-space. Invent. Math. 111 (1993), no. 1, 77–111.

[Sch] Schlenker, J.-M. Dihedral angles of convex polyhedra. Discrete Comput, Geom. 23 (2000), no. 3, 409–417.

[Sto] Stoker, J. J. Geometrical problems concerning polyhedra in the large, Comm. Pure Appl. Math. 21 (1968), 119–168.

[CB] Casson, Andrew J.; Bleiler, Steven A. Automorphisms of surfaces after Nielsen and Thurston. London Mathematical Society Student Texts, 9. Cambridge University Press, Cambridge, 1988.

[Ham] Hamilton, Richard S. The Ricci flow on surfaces. Mathematics and General relativity, 71 (1988), 237–262.

[Cho] Chow, Bennett. The Ricci flow on the 2-sphere, J. Differential Geom. 33 (1991), no. 2, 325–334.

[Gol] Goldman, William. Projective geometry on manifolds. Lecture Notes (1988).

[Mil] Milnor, John W. On the existence of a connection with curvature zero, Comm. Math. Helv. 32 (1958), 215–223.

[Lof] Loftin, John C. Applications of Affine differential geometry to $RP(2)$ surfaces. PhD in mathematics (1999), Harvard University.

[RS] Rodin, Burt.; Sullivan, Dennis. The convergence of circle packings to the Riemann mapping, J. Differential Geom. 26 (1987), 349–360.

[HS] He, Zheng-Xu; Schramm, Oded. On the convergence of circle packings to the Riemann map, Invent. Math. 125 (1996), 285–305.

[St] Stephenson, Kenneth; Introduction to circle packing: The theory of discrete analytic functions. Cambridge University Proess (2005).

[CLN] Chow, Bennett; Lu Peng;Ni Lei. Hamilton's Ricci Flow. Lectures in Contemporary Mathematics, Vol. 3. American Mathematical Society and Science Press (2006).

[CL] Chow Bennett; Luo Feng. Combinatorial ricci flows on surfaces. J. Differential Geom. 63 (2003), no.1, 197–129.

[KSS] Kharevych Liliya; Springborn Boris;Schröder Peter. Discrete conformal mappings via circle patterns. ACM Transactions on Graphics 25 (2006), no. 2, 412–438.

[JLG] Jin M.; Luo F.; Gu X. Computing general geometric structures on surfaces using ricci flow. Computer-Aided Design, 39(2007), no.8, 663–675.

[HJY] Gu X.;He Y.;Jin M.; Luo F.; Qin H.; Yau S.-T. Manifold splines with single extraordinary point. Proc. of the 2007 ACM Sym. Solid and Physical Modeling, 61–72.

[JLYG] Jin M.;Luo F.; Yau S.-T.;Gu X. Computing geodesic spectra of surfaces. Proc. of the 2007 ACM Sym. Solid and Physical Modeling, 387–393.

[FH] Floater, Michael S.; Hormann Kai. Surface parameterization: a tutorial and survey. Springer-Verlag, Heidelberg (2005), Advanced in multiresolution for geometric modeling, editors N.A. Dodgson, M.S. Floater and M.A. Sabin. 157–186.

[SPR] Sheffer, Alla; Praun Emil; Rose Kenneth. Mesh parameterization methods and their applications, vol. 2 of Foundations and trends in computer graphics and vision. Now Publishers Inc. (2006).

Index

高等教育出版社自然科学学术出版中心

高等教育出版社是教育部所属的国内最大的教育出版基地, 其自然科学学术出版中心下设研究生教育与学术著作分社和自然科学学术期刊分社, 正努力成为中国最重要的学术著作出版单位和最大的学术期刊群出版单位.

研究生教育与学术著作分社充分发掘国内外出版资源, 为研究生及高层次读者服务, 已出版《教育部推荐研究生教学用书》、《当代科学前沿论丛》、《中国科学院研究生院教材》、《中国工程院院士文库》、《长江学者论丛》等一系列研究生教材和优秀学术著作.

自然科学学术期刊分社主要负责教育部大型英文系列学术期刊出版项目 Frontiers in China 中基础科学、生命科学、工程技术类期刊的出版工作, 目标是搭建国内学术界与海外交流的平台, 以及国内学术期刊界合作的平台.

地　　址: 北京市朝阳区惠新东街 4 号富盛大厦 15 层 (100029)

网　　址: http://academic.hep.com.cn/

购书电话: 010–58581114/1115/1116/1117/1118